AF841253

THE LITTLE BOOK OF THE MIND

First published in 2012 by New Holland Publishers
London • Cape Town • Sydney • Auckland
www.newhollandpublishers.com

Garfield House
86–88 Edgware Road
London W2 2EA, UK

Wembley Square
Solan Road Gardens
Cape Town 8001
South Africa

Unit 1, 66 Gibbes Street
Chatswood
New South Wales 2067
Australia

218 Lake Road
Northcote
Auckland
New Zealand

10 9 8 7 6 5 4 3 2 1

Copyright © 2012 in text: Surendra Verma
Copyright © 2012 New Holland Publishers Ltd

All rights reserved. No part of this publication may be reproduced, stored in a retrieval system or transmitted, in any form or by any means, electronic, mechanical, photocopying, recording or otherwise, without the prior written permission of the publishers and copyright holders.

ISBN 9781742572130

Publisher: Fiona Schultz
Publishing Director: Lliane Clarke
Project Editor: Simona Hill
Designer: Stephanie Foti
Production Director: Olga Dementiev
Printed by: Ligare Pty Ltd

Other New Holland titles by Surendra Verma
The Little Book of Scientific Principles, Theories & Things
The Little Book of Maths Theorems, Theories & Things
The Little Book of Unscientific Propositions, Theories & Things
Who Killed T. Rex?: Uncover the Mystery of the Missing Dinosaurs

THE LITTLE BOOK OF THE MIND

HOW WE THINK AND WHY WE THINK

NEW HOLLAND

SURENDRA VERMA

Contents

Introduction

You are like a flowering tree,
amazed when I praise you for your gifts.
Rabindranath Tagore, *Fireflies* (1928)

Western philosophy began in the sixth century BC with Thales of Miletus, one of the seven sages of ancient Greece. He was also a keen astronomer and was the first to predict a solar eclipse. One night he was walking along gazing at the sky and fell into a ditch. A clever and pretty girl lifted him out and remarked sarcastically, 'Here's a man who wants to study stars, but cannot see what lies at his feet.' This incident gave birth to the image of the archetypical absent-minded philosopher (or professor).

Though Thales couldn't see what lay at his feet, he was fully aware of what lay within his mind because he gave us the eternal aphorism, Know Thyself. To be wise we must have a better understanding of ourselves: our thoughts, actions, motivations, feelings and emotions.

This book presents 100 insights — not philosophical or spiritual, but strictly scientific — into our minds and emotions. Absent-minded he might have been, but 'Professor' Thales would have agreed with the scientific nature of these insights for he was the founder of Western science.

Stories, to use journalistic jargon, in this book provide glimpses of up-to-date scientific thinking on the matters of mind. New research may contradict some of this thinking. A report on a research study you may read in the popular media doesn't mean the study result has been automatically stamped 'proven by science'. Science advances unpredictably, not linearly in a series of eureka moments; one scientific study often disputes the other, sometimes followed by the third that disputes both. An idea is labelled truly 'scientific' only when it has earned the consensus among the majority of scientists in that particular field.

The stories here are short, sharp and simple and require absolutely no background in science. They are interrelated, yet each is complete in itself. For ease of navigation, they are loosely organised in four sections: (1) The Creative and Competent Mind; (2) The Individual and Interactive Mind; (3) The Motley Mind; and (4) The Mysterious Mind.

It's a book that wants to make you 100 per cent informed about 'gifts' of your mind. Read it from cover to cover or dip into it at random.

The Creative and Competent Mind

1

Bookmark this page

Arthur Fry, an engineer and a keen choir singer, was in the habit of inserting little scraps of paper into his choir book so that he could quickly find the right hymns. One Sunday morning in 1974, when he stood up from the front pews of his church to join the choir, his bookmarks slipped out and fell on the floor. He was embarrassed and frustrated. The singing was followed by the Sunday sermon. Instead of listening to the 'dull sermon' he started to daydream about the bookmark problem. 'It was during the sermon,' Fry remembers, 'that I first thought: What I really need is a little bookmark that could be stuck on, and removed, without damaging the book.'

Then came the eureka moment: it occurred to him that an adhesive invented by his colleague Spencer Silver at 3M in St Paul, Minnesota, could be used to create a better bookmark. Six years earlier Silver had developed an adhesive that stuck readily, but not tightly, to surfaces. As happens with many inventions, Silver couldn't find a marketable use for his adhesive. Returning to work, Fry used some of Silver's adhesive to coat his markers. The markers stayed in place, yet lifted off without damaging the pages. The yellow Post-it note, one of the most popular office products, was born. The humble Post-it note is a monument to the creativity unleashed by a daydream.

Daydreaming — brown study, building castles in the air, reverie, lost in thought, mind-wandering, wool-gathering or whatever you call it — has always been ridiculed as a lazy habit. We daydream for about one-third of our waking hours, although a single daydream lasts only a few minutes. Yet this ubiquitous human experience has been frowned upon as a non-activity in societies that preach productivity: work, achieve, produce. Only slackers daydream, not alpha males and females.

An empty mind is devil's workshop, so goes the old saying. All evil deeds start in the mind; don't sit idle; keep your mind occupied in something positive, something worthwhile — the Jeremiah sermon you must endure. Even if you are resting physically, occupy your mind in healthy reading or mental chanting of God's name — the Maharishi mantra you are advised to follow. But why?

It's not for us to say whether what follows qualifies as *healthy* reading.

Not to worry. Bookmark this page. Sit down, put your feet up, savour a few moments of solitude and daydreams and then read on.

Daydreaming gives our brains an amazing capacity to multitask, which can help solve problems and unleash creativity

2

Wandering mind works wonder

Caught daydreaming at your desk? Don't feel guilty. Blame your brain. In the absence of engaging tasks your brain has simply returned to its normal state. The human brain has been purposefully designed — or hardwired, as they say these days — for daydreaming; it is our minds' default mode of thought. Brain scans show that our brains have a 'default network', a region that remains active when we are daydreaming. When we have a specific task our minds focus on that task. But during familiar tasks, such as making a peanut butter sandwich, when there is no external demand for thought, our minds do not go blank. Instead they tend to wander, moving swiftly from one thought to the next, generating images, voices and feelings. Most of the time the wandering thoughts are not fanciful; they usually are personal thoughts such as working out everyday problems or making plans for the near future.

Psychologists tell us that we spend most of our time engaged in goal-directed thoughts and occasionally we have blips of irrelevant thoughts that pop on the radar. Recent research, however, points otherwise: most of the time we are engaged in less directed, unintended thoughts, and that state is routinely interrupted by periods of goal-directed thoughts. The human brain prefers its default mode. But it immediately springs into action when some task requires attention.

A team led by American cognitive scientist Malina Mason trained volunteers to become proficient in certain verbal and visual memory tasks (for example, remembering and manipulating four-letters sequence such as 'R H V X') so that their minds would be able to wander when they did them, but would have to concentrate when given a new task. The real-time images of volunteers' brains showed that the 'default network' was active when the volunteers were not engaged in a focused task. When the brain was supposedly doing nothing, it was really doing a tremendous amount.

Why do we daydream? Perhaps it keeps our brains active during mundane tasks, or perhaps we daydream because we can. Whatever the reason, daydreaming gives our brains an amazing capacity to multitask. It gives us a sense of 'left-over' resources which the brain can use to work out some problems or anticipate what needs to be done in the future, says Mason. 'Without that skill we'd be pretty limited creatures.'

During daydreaming the brain scan is well lit up, showing that the brain is still using enormous amounts of energy: only 5 per cent less than when it is active. This gem of information gives you an excuse to leave your desk and get a coffee to replenish your body's energy. While walking back to your desk, don't forget to stop at the water cooler for the latest salacious office gossip. Gossiping is considered a form of improvisational daydreaming. Like daydreaming, it is also considered an 'evil' pleasure. Neuroscientists tell us that our brains are wired to respond to gossip, and this activity always fires up imagination. 'Intelligence is imagination with an erection,' Victor Hugo is believed to have said during one of his phallocratic moments, according to the art critic Robert Hughes.

We do not have any evidence of the link between Hugo's daydreaming and his creativity, but now we do have enough experimental proof to say that daydreaming certainly boosts creativity. It's an important tool for creativity. The daydreaming mind is not shackled to its immediate surroundings; it is free to go anywhere; it is free to make new associations and connections; it is free to engage in abstract thought and imaginative ramblings.

The intense focus on a problem has its advantages, but the relaxed style of thinking leads us to contemplate ideas that sometimes seem silly or far-fetched. Such imaginative thoughts might not be practical, but they often are the perfect springboard to creative insights.

Follow the simple exercise, 'capturing a daydream', that American psychologist Robert Epstein has developed to persuade people of their creative potential: 'Close your eyes. Let your mind wander for a few minutes. Relax and just let your thoughts go without deliberately guiding them.'

Daydreaming is relaxation, a kind of meditation, a kind of micro-holiday from which you come back fully recharged and refreshed. You can use the virtual world of daydreams to control your fears and phobias by exploring painful scenarios. For example, if you suffer from claustrophobia — fear of confined places — and dread making trips in elevators, you can daydream ways to suppress this phobia. Try daydreaming the trip with your friends, how you could occupy your mind by talking to them, how you could avoid a panic attack by deep breathing before you enter the lift.

Daydreaming also has social benefits. We tend to daydream about people we love. In a way our daydreams about our loved ones are helping us psychologically to maintain the relationship. In the same way, daydreaming about an argument you had with someone is like pushing the replay button: you look at the way you have behaved and then imagine scenarios about what would have happened if you had behaved differently. Each scenario gives you a better understanding of yourself and enforces your abilities to handle similar situations in the future.

Some of our daydreams might seem like soap operas, but they give us the ability to reflect on social interactions, both real and make-believe. It may seem incredible that *The Young and the Restless* of your daydreams is reinforcing your own sense of self and unconsciously making you a likable lad or lass in your office.

Put this book down, close your eyes and switch on your mind's daydreaming channel. Let your mind wander. Sometimes it's good to be busy doing nothing. 'Sometimes' is the keyword as any couch potato's expanding girth will tell you the flip side of this maxim. And if daydreaming becomes compulsive, it can consume your real life and turn you into a depressive sack of potatoes.

It's also important that you make sure your mind is not in its default mode at the wrong time, especially when you are driving in busy traffic after the office meeting in which your ambitious plan had been dismissed as a daydream by your smirking nemesis. Walking or driving daydreamers are a danger to themselves and others.

Freud's verdict on daydreaming is
harsh: happy people never make fantasies

3
Forget Freud

In matters of mind — especially dreaming and daydreaming mind — Sigmund Freud once reigned supreme. You may have never read a word of Freud's major works — *The Interpretation of Dreams and Three Essays on the Theory of Sexuality* — but you probably know some Freud speak: Oedipus complex, id, ego, superego, sexual sublimation, repressed memories, and so on. And you can't say that you never made a Freudian slip (an unintentional error in speech) which, in today's language, could be explained as a text message or email from your unconscious mind revealing your suppressed thoughts or feelings.

Freudian psychoanalysis is based on the belief that our emotions and behaviour arise from unconscious fears and desires. The past shapes the present, and if we can trace the source of our unconscious fears and desires to their historical origins — often our childhood experiences — we can understand our troubles and deal better with the realities of life. All you have to do is to lie on the therapist's couch and talk about anything that comes to mind — and the source of your current problems would slowly begin to appear.

The source of our daydreams, according to Freud, is our conscious wishes, whereas the source of our nocturnal dreams is our repressed wishes. His verdict on daydreaming is harsh: 'Happy people never make fantasies, only unsatisfied ones do.'

Freud changed the way we see ourselves, but the question remains: is psychoanalysis an objective science or a pseudoscience (like phrenology)? Psychoanalysis has not yet proved itself a science empirically; nor has psychoanalysis yet been proved quackery. 'If often he was wrong and, at times, absurd, to us he is no more a person now but a whole climate of opinion.' This comment from the poet W.H. Auden after Freud's death in 1939 still holds true, and the present climate of opinion is foggy.

In such a climate, it's better to forget Freud and bring some sunshine into your life. Think of the ways in which daydreaming can benefit you.

Daydreaming helps you:

- relieve monotony and boredom
- relax
- multitask
- find solutions to difficult problems
- boost creativity
- manage social conflicts
- maintain relationships.

The best way to tackle a problem
is with an open and relaxed mind

4

Don't think too hard, inside or outside the square

When someone throws the clichéd 'think outside the square' at you, tell them it can only be done if you stop focusing on the problem and start daydreaming. As long as you are focused on the problem, you're within the proverbial square or box. Only when you let your mind wander, can you walk out of it. The best way to tackle a problem is with an open mind. However, it's easy to let your mind wander; the difficult part is to maintain enough awareness to interrupt your daydream when you notice a creative thought.

Most of the time when we solve a problem we follow a smooth path through to the solution. But sometimes our minds hit a mental block which hinders further progress. Restructuring the information allows the problem-solver to clear the mental blockage. A deeper understanding of the problems and its solutions follows and then the sudden flash of insight or the 'Aha!' experience.

The brain generates electrical impulses, or waves, which are known as gamma, beta, alpha, theta and delta waves and show different stages of mind. Simone Sandkühle and Joydeep Bhattacharya of the Austrian Academy of Sciences gave volunteers simple word association problems; for example, find the word that forms a valid compound word or phrase with each of the three words: back, clip, wall (the answer is paper: paperback, paperclip, wallpaper). Such problems involve mental processes similar to processes involved in solving complex problems. The EEG (electroencephalograph) of brain activity of the volunteers showed that the gamma waves were associated with mental blocks.

The higher the focus of the mind, the higher the frequency of gamma waves and less likely the volunteers were to hit upon the answer. The volunteers were provided a clue when they had a mental block. They made better use of the clue when the EEG registered alpha waves associated with a relaxed state of mind.

Simply put, a eureka moment is likely to happen only when your mind is relaxed. The lives of creative people are full of stories showing how daydreaming helped them to create new ideas and insights. Einstein said that his best ideas came while engaged in 'something like' daydreaming. Newton's absent-mindedness and daydreaming are legendary. While working on his monumental theory of gravitation, he used to spend whole days sitting at his bed, half-dressed, thinking and daydreaming.

Everyone knows the story of the first eureka moment in history, the tale of Archimedes running naked through the street shouting 'Eureka! Eureka!' when he had been in the public bath and realised that he could prove whether King Hiero's gold crown was adulterated without damaging it. Daydreaming in the bathhouse, Archimedes watched water being displaced as he got into the bath and worked out the principle of buoyancy and the fact that he could measure the volume of the gold by displacement. He also gave all discoverers a word with which to hail their discoveries and an excuse, if they wish, to run naked through the streets.

Another famous story is that of the Russian chemist Dmitri Mendeleev. In 1869 he was struggling with the problem of the order in which to introduce the sixty-one elements then known in his new textbook of chemistry. He listed the names and properties of the elements on individual cards and began a lengthy game of solitaire (patience), trying to arrange the cards in different ways. Tired, he fell asleep at his desk and dreamed. 'I saw in a dream a table,' he wrote later, 'where all the elements fell into place as required.' The table he dreamed in his reverie became the periodic table, the greatest breakthrough in the history of chemistry.

Perhaps sleeping at your desk is not a bad idea after all. Chemists seemed to have mastered this art. In 1865 Friedrich Kekulé, a German chemist now remembered as the father of structural organic chemistry, also dozed at his desk: 'I was at my desk writing a textbook but my thoughts were elsewhere. I turned my chair to the fire and dozed. The atoms were gambolling before my eyes. I could distinguish long rows of them twisting and twining in snakelike motion. Then one of the snakes grabbed hold of its own tail and began rotating before my eyes.' On awakening he saw the possibility that benzene molecules could be ring-shaped. He had been working on this problem for years but could not imagine that a ring-shaped molecule because all other molecules known at that time had straight chain-like structures.

Kekulé once remarked: 'Let us learn to dream, gentlemen, then perhaps we will find the truth … but let us beware of publishing our dreams before they have been put to the proof by waking understanding.'

Follow the great chemists, doze at your desk and daydream, but maintain enough awareness to capture your daydream and turn it into a working idea.

If you are inspired by Archimedes, you may try your bath, but don't run naked in the street.

The right side of our brains literally lights up
when we experience a sudden 'Aha!' moment

5

Light up that bulb in your brain

Designers often symbolise creativity with a light bulb. This popular design symbol has now the approval of neuroscientists. They have found that a small region on the right side of the brain shows a striking increase in electrical activity (the brain scan literally lights up) when people experience a sudden eureka moment.

Researchers scanned volunteers' brains while doing simple word puzzles. The puzzles involved finding a common word that linked three different words, for example, fence, card and master (answer: post). In more than half the cases the answer came in a flash of insight. The volunteers' 'Aha!' moment generated a striking increase in electrical activity in the right side of the brain.

Well, what could you do to light up that bulb in your brain?

American psychologist Robert Epstein believes in four core competencies of creative expression:

Capturing. Whenever you get new ideas, learn to preserve them.
Surrounding. Surround yourself with interesting people and things.
Challenging. Tackle tough problems.
Broadening. Expand your knowledge.

Once you have acquired these four core competencies, the following tips, based on a list devised by Ulrich Kraft, a German physician, would help you to be creative:

Wonderment. Retain a child-like curiosity about the world around you.

Motivation. You know Thomas Alva Edison's famous aphorism: Genius is 1 per cent inspiration and 99 per cent perspiration. As soon as the first spark of 1 per cent inspiration arises, follow it with determination.

Intellectual courage. Think outside the square. Avoid using this over-used phrase in your reports and talks, but practise it.

Relaxation. Relax, ponder, daydream.

Laughter gives our brains a rush
of endorphins, which reduces stress

6

Laugh and be creative

The simplest and the fastest way to relax is to laugh. The old saying 'laughter is the best medicine' is indeed true. Laughter offsets the impact of mental stress and relaxes your heart — and mind. If the following joke has made you laugh, then you will find it easier to light up that bulb in your brain:

> The bazaars of India are famous for haggling. The locals taught a foreign visitor the best bargaining trick: just reduce the asking price by half. When a stallholder asked 200 rupees for a shirt, she offered 100 rupees. The stallholder replied that he would accept 150 rupees, nothing less. The visitor said: 'I'll give you 75 rupees only.' 'Okay, you take it for 100 rupees,' the stallholder relented. '50 rupees, that's all you will get,' the visitor stressed. The exasperated stallholder became angry and said: 'You take it for free!' 'No, I'll take two,' the well-trained shopper replied.

Laugh loudly, even if you have to force yourself. Keep laughing. Laughter gives our brains a rush of endorphins, the hormones that are responsible for our general sense of wellbeing. After just five minutes of laughter you'll start feeling really great.

Can't laugh because you have been reading this book at your desk in office? Just smile. They say it takes seventy-two muscle to frown and only fourteen muscles to smile. These fourteen muscles also work very hard to release some of the same feel-good endorphins.

Don't want to smile because there is no one to smile at? Okay, just eat a small bar of dark chocolate. Chocolate is good for the brain; it also triggers the release of endorphins but not as effectively as laughter.

Your relaxed brain is now ready to solve the problem you have been mulling over for a long time. Aha!

You don't have to be a Michelangelo, Mozart or Einstein to be creative. You can express your creativity in everyday life: devising a new recipe, landscaping your garden, painting, handicrafts ... the list is endless.

To be creative:

- Know your stuff. Immerse yourself in the problem. Ask questions.
- Keep trying. You are unlikely to solve a big problem at the first try, or the second, or the third ...
- Be ready to imagine the impossible. Many breakthroughs at first seem completely crazy.
- Relax: laugh, daydream, doze.

Meditation, deep breaths or
a nap can increase brain power

7

Say Om to boost creativity, or join the midday snooze club

Meditation — of any kind, not necessarily chanting Om or other mantra with a saffron-robed guru — fosters inner peace and relaxation. This claim has been supported by decades of research showing changes in brainwave patterns during meditation.

In our brains, a large mass of grey matter called thalamus acts as the gatekeeper by relaying sensory information. It focuses our attention by funnelling data into the brain and stopping other signals in their tracks. Brain scans during meditation show that the flowing of incoming information in the thalamus reduces to a trickle. This is a sign that meditation has not shut off the brain but rather it has blocked information from coming into the part of the brain responsible for processing it. There is also a decrease in beta waves, waves associated with a fully awake mind. At the same time, alpha and theta waves, waves associated with a relaxed mind, are extremely active. In long-term meditators, theta waves dominate the brain during periods of deep relaxation.

Alpha waves are associated with daydreaming, alpha and theta waves with meditation. The ideal state of consciousness for creativity happens when the mind is flooded with a mixture of alpha and theta waves. If your neurons have made the right connections, you know now that to be creative you have to find the right balance between daydreaming and focused thinking, or between fantasy and reality. Daydreaming can help you to relax; meditation can help you to cultivate concentration by not thinking.

Buddhist monks say that our mind is like a deer making its way through a thick forest, its antlers getting caught in branches time after time. Meditation can turn it into a monkey, swinging effortlessly from branch to branch. Some psychologists believe that information is digested into our subconscious before it is turned into ideas. Real insights happen when ideas from the subconscious mind can effortlessly jump like a monkey to the conscious mind.

The meditating mind is not an empty mind. We do not 'empty' our head during meditation, rather we try to train our neurons to direct activity in the concentration-oriented area of the brain. In other words, we train our minds to get used to learning to be totally aware of the moment.

How can you train your mind to strike a balance between awareness and distraction? Simply by meditating daily.

Try this ancient Vedic technique. You do not have to sit cross-legged on the floor. Just sit comfortably in a chair or on the floor. You may even lie down on the floor, but keep your back straight and body relaxed. Shut your eyes. Breathe normally.

In this technique you are required to silently repeat a word, any word. The word, or mantra as it is called in the Vedic literature, has no religious connotation; you don't need a bearded guru to give it to you (after charging a hefty fee). Try the word 'one' or the Sanskrit 'Om'.

After about half a minute of normal breathing, think the word and then let it go. Do not say the word over and over again; do not chant it. Do not focus on the word; do not try to say it. Just think about it. The word is just a faint idea. The word may become louder, softer or fainter. It may speed up or slow down. Just take it as comes.

The word will not make your mind empty. Thoughts will come and go. Do not repress them or push them away. Let them simply pass through your mind. Do not use the word to push them out of your mind. Eventually thoughts may not come at all. If you become aware that your mind is not thinking the word, quietly come back to the word.

If this word or mantra business sounds too silly or too difficult, try this simple version of Buddhist meditation which begins by focusing on your breath. As suggested before, sit comfortably in a chair or the floor. Take in a slow deep breath. Pause for a few seconds then breathe out gently. Observe the entire course of your breathing and let it settle to its natural flow.

Thoughts will come and go. Do not force your attention on to breathing. Quietly return to it. If you hear a noise, just listen to it rather than thinking about it. The idea is to pay attention to sensory experience; not to think about it. The goal of any type of mediation is getting used to not thinking.

Meditation should last at least 20 minutes. Try it twice a day, if possible. After a few sessions, you will start noticing the calming effect on your mind. A calm mind is a prerequisite for creativity.

If you think that you are not cut out to be a meditating type, there are other methods to relax.

Learn to take deep breaths. Take in a slow deep breath. Pause for a few seconds and then breathe out gently. Try it ten times. Try it a few times every day.

Or take a nap. After lunch, forget caffeine, take your shoes off, sit down comfortably and take a 30- to 40-minute nap. Studies show afternoon napping not only improves health and work performance, it also increases brain power. It's time you joined the midday snooze club. You will be in the company of Leonardo da Vinci, Napoleon Bonaparte, Johannes Brahms, Albert Einstein, John F. Kennedy, Bill Clinton and other who have been caught catnapping during their work hours. Sleeping on the job is good for you.

Our minds (infinite and ethereal) seem very different from our brains (finite and material), but it is now generally accepted that mind and brain are really one

8

Your mind is what your brain does

In his early years, the great seventeenth-century French mathematician and philosopher René Descartes, a small timid man with a large head, low forehead and discreet, stubborn and fanciful eyes who spoke in a feeble voice, was sceptical of almost everything, even his own existence. He lost this scepticism after reaching the conclusion, '*Cogito, ergo sum*' ('I think, therefore I am'), philosophy's most famous statement.

When he was sent to a boarding school at the age of eight, he enjoyed exceptional privileges because of his poor health. 'My philosopher,' as his father used to call him, was excused from morning school duties and was allowed to stay in bed until late in the morning. This habit of morning reflections in bed clung to him throughout his life.

In his later years, he reflected upon how to arrive at knowledge without any fear of error and concluded that only two mental acts, intuition and deduction, are the most certain paths to knowledge. Intuition, he said, is the undoubting conception of a pure and attentive mind. Intuition proceeds by deduction, by which we understand all that is necessarily concluded from certain other facts already known.

To discover what, if anything, he can know with certainty, Descartes presents a series of arguments to cast doubt on knowledge he has accepted as truth:

> All those things that have entered my mind were no truer than the illusion of my dreams. Or, a deceiving God or an evil demon is causing me to go wrong about knowledge, which is self-evident and which I seem to see so clearly. Because my

> senses may sometimes deceive me, the source of my knowledge cannot lie in my senses. I can only be certain about the experience of thinking. I cannot doubt that I think, therefore, I exist as long as I'm thinking.

This statement exemplifies his reasoning that he could doubt the physical world but not his mind.

Descartes believed that the mind, something immaterial that holds the essence of a human being, was separate from the brain but interacting with it in some way. He said that mind and body must be fundamentally different: the body was made of physical substance and occupied space whereas the mind was nonphysical and didn't occupy any space. His distinction between the physical and the mental — now known as Cartesian dualism — was sanctioned by the Roman Catholic Church and therefore dominated Western thinking for centuries. It still has a major influence today. Many Eastern mystical traditions, however, teach that the mind and body belong to an individual continuum.

The human brain is gradually yielding its secrets to increasingly powerful tools that can image what happens inside a person's head, say, when doing mathematics or listening to Mozart.

There are more than 100 billion neurons, or nerve cells, in the adult human brain, and each neuron is specialised and connects with up to 10,000 neighbours. Neurons transmit electrical signals from one to another. These signals are carried by molecules across contact points, called the synapses, with other neurons. The molecules are called neurotransmitters — you probably have heard of some of them: dopamine, endorphin, histamine and serotonin. Brain activity is basically just a bunch of neurons firing. When one neuron fires up, it excites its neighbours and they in turn fire up others, giving rise to patterns of activity that result in thoughts, feelings and perceptions.

These days neuroscientists use the sophisticated functional magnetic resonance imaging (fMRI) technique to study brain activity. In this technique a person's brain is scanned in a doughnut-shaped machine while he or she is doing different mental tasks, giving neuroscientists the opportunity to see real-time images.

Neuroscientists can now 'video' the trails of neurons blazed by flitting thoughts and feeling. These 'videos' are now revealing the inner workings of the brain and showing that our thoughts, feelings and perceptions are simply the result of complex electrochemical interactions within and between neurons. So far, neuroscientists have managed to re-create entire video clips just by analysing the brain patterns of people watching them. In other words, as you watch a video your brain is monitored with the help of an fMRI machine. Neuroscientists then reconstruct approximate images of the video from the brain-imaging data. In the near future, instead of plucking pictures from our brains, scientists would be able to decode our thoughts by mapping the brain's response to images, words and emotions.

Already, neuroscientists have recorded the responses of individual neurons. In one experiment they traced a single neuron that fires only when the subject is shown pictures of the former US president Bill Clinton and no one else. Interestingly, in another experiment they found a neuron that fires only when the subject is shown pictures of the actress Jennifer Aniston, but not pictures of her with her former husband, the actor Brad Pitt. Obviously, there are neurons in our brains which fire only when you see someone you know or recognise.

Undoubtedly, mental activity is a type of brain activity. Or, the mind is like software in the brain's hardware. But not everyone agrees with such mechanistic explanations of our mind. The mind–brain debate continues.

Locating the precise brain activity
that creates specific behaviour responses

9

It all began with Tan's trouble

Pierre Broca was a neurologist working in Paris. In 1861 the 21-year-old Monsieur Leborgne was admitted to his hospital. The unfortunate young man was affected by epilepsy and had lost the ability to speak. He understood speech perfectly well, but could only say 'tan' whenever asked a question. For this reason, he was nicknamed Tan.

He died a few days later. When Broca performed an autopsy, he discovered a lesion in the left side of Tan's brain. He concluded that Tan's speech disorder was due to a lesion of the left hemisphere. After further research he named this speech disorder *aphémie* (which was later renamed aphasia).

Within a decade Karl Wernicke, a German neurologist, identified a second area of the brain, near the area identified by Broca, where damage causes another kind of speech disorder. Such patients can speak fluently, but what they say is meaningless; they can't understand speech and therefore they're unaware that they're talking nonsense. These two different kinds of language disorders are named after their discoverers: Broca's aphasia and Wernicke's aphasia. And if you suffer from 'whatsisname' condition (meaning you forget the names of people), it's called onomastic aphasia.

Broca once complained in a letter that 'private practice and marriage' were the twin extinguishers of science, yet his scientific work started the search for what areas of the brain did what. Locating the precise brain activity that creates specific behaviour responses is now an advanced science.

Broca and Wernicke each connected an area of the brain to an element of speech, but by the end of the 20th century scientists had learned that there are still more regions involved in speech. New brain-imaging techniques have allowed scientists to picture the brain during different tasks. The brain has regions with specialised functions, but scientists are not sure whether a specific function is always carried out in the same brain region for everyone. The whole is more than the part, and different brain regions work together like the instruments in an orchestra. But scientists do not yet know how all the different brain regions involved in one task come together.

Today's brain images are a far cry from the head maps of phrenology (*page 171*).

10
Teaching an old dog new tricks

It's not a fairytale, but once upon a time it was thought that the adult brain was an immutable organ in which damaged cells could not be repaired or replaced. The skin, blood, heart, lungs, kidney and liver all could do it, but not the brain. It was a black box. But new brain-imaging techniques have given scientists an unprecedented ability to peer directly into the inner workings of the brain. They have now found evidence that the brain can change even in adulthood, and sometimes these changes can take place within seconds.

A neuron consists of a central cell body and a series of branches that extend into different parts of the brain to fire electrochemical signals from one another. If certain neural pathways are blocked, new pathways open up. As these new pathways continue to be used they become stronger and more prominent — a phenomenon captured in the oft-repeated phrase 'neurons that fire together, wire together'. Neuroplasticity is the ability of our brains to produce new connections between neurons as a result of new experiences. In other words, neuroplasticity is how the brain changes, or rewires, with learning and experiences. The brain can also produce new neurons — not only in children but also in older people. This process is called neurogenesis.

A classic study of London taxi drivers skilled at navigating complicated city routes provides an example of neuroplasticity. The taxi drivers had larger hippocampuses than London bus drivers who followed a fixed route. Moreover, the longer these taxi drivers drove, the larger was the size of the hippocampus, the brain's centre of learning and memory.

Other studies have shown that a greater use of a particular muscle causes the motor cortex, the brain's region that controls voluntary body movements, to grow. In a pioneering experiment, Harvard Medical School neuroscientist Alvaro Pascual-Leone compared the brain scans of two groups of volunteers: those who practised piano exercises for a week, and those who merely thought about practising the piano exercises (holding their hands still while imagining how they would move their fingers). The physical structure and function of the motor cortex improved in both groups, proving that mental training can change our plastic brain.

This means we have some control over plasticity of our old brains: learning new skills. Learning involves strengthening connections between neurons. This can happen in two ways: by creating more connections between neurons, and by increasing their ability to send signals.

'Spending all middle age reading the newspaper, watching television or dancing the old dances will not give your brain enough stimulation,' advises Norman Doidge, a Canadian psychiatrist and author of bestselling book, *The Brain That Changes Itself*. 'You need novelty. You've got to learn new dances. You've got to tax it — the way you might've had to tax your brain when you were trying to learn a French vocabulary if you're an English speaker in high school.'

To tax your brain you have to learn new 'tricks'. New 'tricks' for 'old dogs' also include aerobic exercise or walking — a growing number of studies shows that physical exercise improves memory and attention, which decline in old age. Another 'trick' is meditation.

11

Flexing muscles in the brain

Research using brain-imaging techniques suggests that meditation is doing more than changing wave patterns in the brain (*page 26*); it may even be rewiring it. Brain scans of long-term meditators show changes in the actual structure of the cerebral cortex, the outer parts of the brain usually referred to as the grey matter or thinking cap. Associated with reasoning and sensory perception, these parts have been found to be thicker in meditators than non-meditators. The thickening was more pronounced in older than in younger people. This has intrigued researchers because these parts of the cerebral cortex normally shrink as we age. Those deeply involved in meditation show the greatest thickening, confirming that it was caused by extensive practice.

People who meditate not only show more grey matter, they also have stronger connections between different regions. These connections increase the ability of neurons to rapidly relay signals in the brain. Studies have shown the increase to be throughout the brain, not just in certain areas.

To discount the idea that people with a thicker cerebral cortex were more drawn to meditation, researchers studied the brain of volunteers before and after eight weeks of meditation sessions, which ran on average for 27 minutes per day. They found increased grey matter in the hippocampus, a continuation of the cerebral cortex (known to be important for learning and memory), and superior frontal gyros, a structure behind the forehead associated with self-awareness, compassion and introspection. This research failed to find any changes in insula, a region within the cerebral cortex, which had been identified in other studies. It's believed that changes in insula take place after meditating for a longer period. Other studies have shown that amygdala is less active in people who mediate. This almond-shaped region of the brain registers fear and stress.

These studies remind us that the brain is plastic and meditation can play an active role in shaping its structural changes to enhance our wellbeing and quality of life.

Brains' overall size and
weight are not related to intelligence

12

Are brilliant brains bigger?

When Carl Friedrich Gauss (the great German mathematician who can be ranked only with Archimedes and Isaac Newton, 'and it is not for ordinary mortals to attempt to range them in order of merit,' warns E.T. Bell, the famous historian of mathematics) died in 1855, a Dr Rudolph Wagner compared his brain with the brain of an ordinary day labourer and found it to be equal in all respects.

After Albert Einstein died in his sleep in 1955 the pathologist Thomas Harvey performed an autopsy. During the procedure he removed the brain and instead of putting it back in the skull he put it in a glass jar of formaldehyde. His excuse was 'a sense of duty to science'. Harvey had no training in neuroscience but kept the jar for decades for 'scientific study'. He never published any research findings, but after three decades he doled out small sections of the brain to a few neuroscientists.

An analysis in 1985 at the University of California showed that Einstein's brain didn't have more neurons than a normal brain, but it had more glial (which means glue) cells. At the time scientists believed that glia were simply cellular packaging to support the intricate network of neurons which do the important work in the brain. Glia are now believed to be involved in memory and learning; unlike neurons which communicate with each other by electrical pulses, glia communicate with each other in a secret language of chemicals. Did the abundance of 'talking' glia made Einstein smart? We can't be sure because the research on glia is still not conclusive.

In 1999 neuroscientists at McMaster University in Canada, who maintain a 'brain bank' for comparative studies of brain structure and function, also examined some sample tissues of Einstein's brain. Their study revealed a distinctive physical characteristic: the region of the brain that is involved in mathematical thinking and imagery — two key features of the kind of thinking Einstein did best — was about 15 per cent wider than normal, making it spherical.

Other studies have shown that the genius's brain weighed 1.23 kilograms (2 pounds $11^{1}/_{3}$ ounces), slightly less than the average human adult brain, which weighs about 1.4 kilograms (3 pounds). But its neurons were tightly packed, which might have allowed for more interconnections resulting in faster mental processing.

Was Einstein's brain built for brilliance or his thinking about the universe changed his brain? We still do not know what exactly makes someone a genius, but we do know that the brain's overall size and weight are not related to intelligence.

We do not use only 10 per cent of our brains, yet this myth has motivated many people to strive for greater creativity and productivity

13

A comforting myth

You have probably heard that we use only 10 per cent of our brains. Or, on average, that we remember 20 per cent (or some other small percentage) of what we read, 30 per cent of what we hear, 40 per cent of what we see, 50 per cent of what we say, 60 per cent of what we do, and 90 per cent of what we see, hear, say and do. There is absolutely no scientific evidence, even of moderate quality, to support these absurd claims.

Neurons are the basic working units of the brain. They transmit information to other neurons, muscle or gland cells, and they control body movement, perception, emotion and thought. At any given time, not all neurons are active, but no neuroscientist has ever found that 90 per cent of our brain is perpetually on vacation. In recent years neuroscientists have scanned the brain with sophisticated big-name machines such as electroencephalography (EEG), magnetoencephalography (MEG), computerised axial tomography (CAT) and positron emission tomography (PET) and functional magnetic resonance imaging (fMRI) and have pinpointed numerous psychological functions to its specific parts. Their scans have not revealed any portions of the brain that are in a vegetative state. Of course, some parts of the brain are more active than others at any given time or during a particular activity. Besides, if 90 per cent of our brains was really doing nothing, there would be large areas of dead cells in our brains. No autopsy has ever revealed that to be true.

Yet, the myth of the 10 per cent brain refuses to die.

For our body, the brain is an expensive organ to maintain; it uses too many resources:

about 20 per cent of our body's daily calorie intake. Evolution would not have allowed such a wasteful organ to survive.

When someone tells you that we use only 10 per cent of our brains, they are probably using only 10 per cent of their brains. Where does this myth come from? No one knows for sure. The myth has been popular for nearly a century. It has been a boon to self-improvement crusaders. Probably they would have to invent it if it didn't exist. A whole industry is based on this myth.

To be fair, the myth has an uplifting advantage. 'The 10 per cent myth has undoubtedly motivated many people to strive for greater creativity and productivity in their lives — hardly a bad thing,' observes Barry L. Beyerstein, an American psychologist. 'The comfort, encouragement and hope that it has engendered helps to explain its longevity.'

Does multitasking really make us
perform better? Or does it muddle our minds?

14
The myth of multitasking

In good ol' days, multitasking was simply doing two tasks at the same time, such as listening to the radio while driving or reading a newspaper while eating breakfast. In today's fast-paced world, it's not surprising to see someone checking text messages on a mobile phone, sipping a can of drink, finding the sender's number in the contact list and then calling — all this at a red light in a car he or she is driving. Illegal and dangerous this multitasking may be, but is it really more productive?

Our first glimpse into the multitasking brain came in 1935 when the American experimental psychologist John Ridley Stroop discovered that people get frustrated when asked to name aloud the colours of words printed in incompatible ink colours (for example, to say 'red' in response to the word 'green' printed in red ink, and so on). However, when asked simply to read the same list of words (say 'red' in response to the word 'red' printed in green ink, and so on), they breezed through the task. This mental test — now known as the Stroop effect — shows how your brain deals with conflicting information. Your brain can perform automated processes quickly and unconsciously, but when it has to process information for more than one task simultaneously, it manages its limited attention resources by inhibiting or stopping one response in order to say or do something else.

Recent research has given us a better insight into the brain when it has to juggle many tasks. Experiments by Russell Poldrack and his colleagues at the University of California show that multitasking adversely affects how we learn and, as a result, we do not learn well when we are distracted. Focused learners and multitaskers use different parts of the brain when they learn the same thing. Focused learners fire up their brain's hippocampus, a sea-horse-shaped structure that plays critical roles in processing, storing and recalling information. Multitaskers rely on their striatum, the region of the brain that controls our ability to learn new things (it's damaged in patients with Parkinson's disease who have trouble learning new motor skills but can easily remember the past). The researchers note that their research doesn't say never to multitask, just don't multitask while you are trying to learn something new that you hope to remember.

Another experiment by a team at the University of Michigan led by David E. Meyer has demonstrated that multitasking may actually reduce productivity because multitasking causes all the lights in the brain to go dim because there just isn't enough power to go around. The researchers asked participants to write a report and check their email at the same time. Those who jumped back and forth between the tasks took about one-and-a-half times as long as those who concentrated on one task before moving on to another. 'Multitasking is going to slow you down, increasing the chances of mistakes,' says Meyer. 'Disruptions and interruptions are a bad deal from the standpoint of our ability to process information.'

Why are we lousy multitaskers? The answer probably lies in the prefrontal cortex. This 'decision-making' region of the brain is involved in multitasking but its exact role is not yet fully known. When people juggle two assignments, their prefrontal cortex appears to deal with the task one by one, instead of parallel processing as in sensory and motor parts of the brain. This slow processing causes the bottleneck at the central stage of decision-making, which limits our ability to multitask. Some researchers suggest that the speed of processing through the prefrontal cortex can be drastically increased through training and practice.

Multitasking is bad for the brain as it puts stress on it. This is the warning from Stanford University researchers Clifford Nass and his colleagues. They surveyed 262 students on their media consumption habits. The nineteen students who multitasked most (simultaneously reading email, surfing the web, talking on the phone, watching TV, and so on) and twenty-two who multitasked least then took a series of tests. In every test students who multitasked most performed worst. Researchers realised that multitaskers were paying a high mental price because single-minded attention is crucial to learning. Multitasking while trying to learn something new was also affecting their ability to apply what they had learned. 'Heavy multitaskers are often extremely confident in their abilities,' he says. 'But there's evidence that those people are worse at multitasking than most people ... They are suckers for irrelevancy, everything distracts them.' He worries that media multitasking might actually be destroying students' capacity for reasoning.

Message to master multitaskers: by doing more you do not necessarily accomplish more; and you'll get worse at multitasking as you age (our cognitive abilities peak in our 20s and then a slow decline starts). And to all drivers: there is irrefutable evidence that talking on the phone while driving badly impairs the ability to drive.

The internet has changed the way we remember information

15

Google for smarter memory

If you're not into psychology, it's most likely you would never have heard of the term 'transactive memory'. Never mind, the internet is now literally in the palm of your hand, simply Google the phrase and you will find numerous web pages that explain its meaning. Or, press Wikipedia into service (remarks a wit, 'When Wikipedia has a server outage, my apparent IQ drops by 30 points'). The internet has become a transactive memory that we depend upon to remember for us.

Transactive memory is an idea that we rely on our family, friends and work colleagues to coordinate memory for us by making them responsible for certain kinds of information. For example, couples in close relationships rely on each other to act as one another's memory bank, freeing both from duplicating certain memories in their own brains — a wife interested in watching football might not bother to remember football facts because she can ask her husband who knows these facts whenever she wants to know something, while he might rely on her to remember important social dates and birthdays of relatives and friends.

Are we becoming dependent on the internet in the same way that we are dependent on our family and friends for shared memory? To answer this question, Betsy Sparrow, a psychologist at Columbia University in New York, and her colleagues staged a series of experiments to catch participants in the act of relying on future access to information rather than memorising the information themselves.

In the first experiment, Barrow used the Stroop effect (*previous story*) to test whether participants — 160 Harvard University undergraduates working on desktop computers — were inclined to think about the internet when presented with a series of difficult true-or-false trivia questions. Immediately after the questions were posed, coloured words would appear on the screen. When the coloured words were internet-related such as Google and Yahoo, the students answered more slowly showing that they were considering going online for answers.

In the following experiment, the trivia questions were turned into statements. Students were asked to type 40 different trivia statements (for example, 'an ostrich's eye is bigger than its brain') into the computer. Half were told the statements would be saved on the computer and the other half were told they would be erased. Then all the students were asked to write down the statements from memory. Students didn't remember the statements as well when they knew that they would be saved because they thought they could later look them up. But those who were told that their notes would be erased had a better memory of the statements, as if their brains had made an emergency backup.

In another experiment, students were given trivia statements and the names of the folder s in which they were stored. They were better at remembering where information was stored than the information itself.

These experiments suggest that no longer do we make costly efforts to find the things we want; we know the computer or the internet will do it for us. Our brains are adapting to new communications technologies.

'I don't think Google has made us stupid — we're just changing the way we're remembering things,' says Sparrow. 'The experience of losing our internet connection becomes more and more like losing a friend. We must remained plugged in to know what Google knows.'

Benjamin Franklin once said that a man has three great friends: an old wife, an old dog and compound interest. It's time we added a new friend to Franklin's old list: Google.

In the absence of a 'memory pill' — though there may be one in the future — take the following 'memory tips', one a day

16

Making the most of memory

Memory is about making connections. When a memory forms in the brain, it involves a change in the strength of connections between neurons. In the case of short-term memories (last about 30 seconds or less), the change in strength is temporary. When new memories are stabilised (long-term memories), the connections become permanently strengthened. When these new connections let the same neurons fire together later on, we recall those memories. Sounds simple, doesn't? Now let your neurons fire together. Ready, steady, learn ...

Make connections. 'Think about relevant connections between what you wish to learn and what you already know,' advises Anthony J. Greene of the University of Wisconsin. 'Build as many meaningful connections as you can.' It's easier to hang a piece of information on to something you already know.

Practise recall. The best way to consolidate new information in your memory is practising recall, instead of reading or reviewing your notes. Recall drill even beats concept mapping used for learning complicated material. This advice, especially for students, comes from Jeffrey D. Karpicke of Purdue University in Indiana. 'When students have the material right in front of them, they think they know it better than they actually do,' he says. 'Many students do not realise that putting the material away and practising retrieval is such a potent study strategy.' Repeated recall is good for memory.

Say it aloud. If you're trying to learn a short piece of information, say it aloud and listen to yourself saying it. This strategy works very well for older people who may have 'tip of the tongue' memory lapses, such as forgetting people's names. Say the new name or word out loud a few times and you're likely to recall it easily.

Don't rush. Learning information over several short sessions is better than doing so in a single long session. Pauses between sessions help solidify new information. Spread out your learning.

Paint a picture in your mind. Visualise new information. You are likely to forget boring facts but not vivid and interesting experiences.

Forget it. Forgetting is good for your memory. People who are good at forgetting unnecessary information are also good at problem-solving and remembering important things even when they are distracted. That's the conclusion of research by psychologist Ben Storm of the University of Illinois. 'We need to be able to update our memory so we can remember and think about the things that are currently relevant,' he says. In Robert Louis Stevenson's 1886 novel *Kidnapped*, Alan Breck knew this secret when he said, 'I've a grand memory for forgetting, David'.

Other helpful memory menders and minders are: meditation, sleep, learning a new language and physical exercise. If you're a worrier, focus on your worries by writing about them, especially before a big exam or performance. The writing will soothe your mind (*page 51*). And don't bother about brain-training programs (*page 73*).

Sleep boosts memory, but
lack of sleep results in poor memor

17

Sleep on it

If you want a memory like an elephant, you're advised to follow the old adage: 'sleep on it'.

Another old saying that elephants never forget is also backed by science. Elephants build up a memory over years and hold on to it. This long-term memory helps them to survive in the wild. For example, when a severe drought hit a Tanzanian national park in 1993, elephants that left the park were better able to survive than those that stayed in the park. These elephants were old enough to remember an earlier severe drought from 1958 to 1961; the elephants that stayed behind were younger.

We don't know whether a good night's sleep helps elephants remember better, but in humans sleep not only helps retain newly learned information but may also assist in recalling it when you need it.

A new piece of information such as a new phone number or a new name would be quickly forgotten unless it becomes a permanent record in our long-term memory. Conversion from short-term memory (lasting less than 30 seconds) to long-term memory (lasting longer than 30 seconds) is called 'memory consolidation'. It occurs when connections between neurons as well as different brain regions are strengthened. Changes in neurons typically take place within the first minutes or hours of learning. Other changes, such as the reorganisation of neuron networks that handle individual memories, can take several days or years.

In a recent study, two groups of volunteers learned new spoken words (for example, 'cathedruke') that overlapped with familiar words ('cathedral'). The first group of volunteers learned the words in the evening, followed by an immediate test. They slept overnight in the laboratory while their brain activity was recorded using an electroencephalogram (EEG) and were given a second test in the morning. The second group of volunteers learned the words in the morning and were retested in the evening, with no sleep in between. Volunteers in the sleep group recalled more words and recognised them faster after sleep. An examination of their brainwaves showed that deep sleep (slow-wave sleep) rather than light sleep (rapid-eye-movement, REM, sleep) helped in consolidating new memories. 'New memories are only really useful if you can connect them to information you already know,' says Jakke Tamminen of the University of York, the lead researcher of the study. A night of sleep enhances new memories by strengthening and integrating them with existing memories.

Many other studies also support the idea that learning continues in sleep. Sleep not only makes memories stronger, it also reorganises and restructures memories to help you produce new and creative ideas. It's a misunderstanding that the sleeping brain isn't doing anything: it's busy organising memories and picking out the most important information — making you come up with new ideas.

A sleepless night, on the other hand, results in poor subsequent retention of information. Staying up all night to cram for an exam may work for you, but most likely you will forget what you have learned in a few days.

As 'sleep on it' has now earned science's elephant stamp of approval, it's time you practised this astute adage — and note that although a nightly sleep is important for learning, even a short midday nap helps ... zzzzz.

Health benefits of writing
about personal experiences

18

The writing cure

Patients who write about their traumatic or emotional experiences show significant improvement in their physical and emotional health. This no-risk treatment may sound like fiction, but a large body of evidence shows that it's a useful supplement to regular medication. Many studies also show that writing about positive experiences can provide health benefits as well.

The link between writing about emotional upheavals and health benefits was discovered in the 1980s by American psychologist James W. Pennebaker and one of his graduate students, Sandra Beall. In the first study of its kind, they asked college students to write continuously for 15 minutes on four consecutive days about the most upsetting or traumatic experience of their lives: these students wrote about the divorce of parents, about loss and abuse, about alcoholism and suicide attempts. A control group of students wrote about superficial topics such as their rooms or clothes. Four months later, the students who wrote about their traumatic experiences showed significant benefits in physical health. 'Apparently, putting emotional experiences into language changed the ways people thought about their upheavals,' Pennebaker says.

Since this pioneering study, numerous other studies have demonstrated that asthma, rheumatoid arthritis, cancer, heart and HIV patients who write about stressful life experiences for a few minutes each day for a few consecutive days show surprisingly beneficial effects that can last for months.

In his book, *Writing to Heal: A Guided Journal for Recovering from Trauma and Emotional*

Upheaval, and his general research, Pennebaker recommends the following basic assignment:

> Over the next four days, write about your deepest emotions and thoughts about the emotional upheaval that has been influencing your life the most. In your writing, really let go and explore the event and how it has affected you. You might tie this experience to your childhood, your relationship with your parents, people you have loved or love now, or even your career. Write continuously for 20 minutes.

You can write by hand or on a computer — any time but not at bed time. Turn on some soft music, think of a topic and then start writing freely without worrying about spelling or grammar. Pennebaker warns that after writing you may feel somewhat sad or depressed. 'Like seeing a sad movie, this typically goes away in a couple of hours,' he says. Whether you keep your writing samples is really up to you, but he offers other options: 'Burn them. Erase them. Shred them. Flush them. Tear them into little pieces and toss them into the ocean or let the wind take them away. Eat them (not recommended).'

If you don't like the idea of solitary writing and want to share your thoughts with other writers in similar situations, blogging is the way to go. Blogging offers the additional advantage of connecting with others and witnessing each other's stories as they unfold. You are no longer alone; you're part of a community.

You don't have to be a patient to reap the health benefits of writing about personal experiences, thoughts and feelings. Evidence is mounting that writing about your intensely positive experiences or on an emotionally positive topic or about your best future self can help improve your psychological wellbeing and physical health. For older people, writing their memories by themselves also shows benefits for their psychological wellbeing because it helps in getting rid of unhappy memories or strong emotions and providing a meaningful purpose.

19

Mixing up letters — and sounds

On 7 November 1896, the prestigious *British Medical Journal* published an article called 'A case of congenital word blindness' by Dr W. Pringle Morgan in which the author described the paradoxical case of a bright and intelligent boy, Percy F., aged 14: 'His great difficulty has been — and is now — his ability to read ... In writing his own name he made a mistake, putting "Precy" for "Percy" ... He seems to have no power of preserving and storing up the visual impression produced by words — hence words, though seen, have no significance for him.' Morgan ended the article on a positive note: 'by dint of constant application this defect has been overcome.'

This is the first medical description of the reading disorder we know now as dyslexia. The paradox of dyslexia —difficulty in learning to read despite normal intelligence — still intrigues scientists after more than a century, but neuroscience is now providing new insights into this lifelong disorder that can be treated but not cured.

We no longer think of dyslexia as simply a visual disorder. People with dyslexia not only have a hard time with written words, they also find it much harder to connect the sounds of words with their meanings. American cognitive neuroscientist John Gabrieli calls dyslexia 'a phonological difficulty in perceiving the sounds that make up words.' He believes it may be more accurate to think of it as a language disability than simply a visual disorder.

To test the hypothesis that dyslexia impairs voice recognition, Gabrieli asked college students with and without dyslexia to listen to recorded voices paired with unique cartoon avatars on computer screens. The participants tried matching voices to the correct avatars speaking in English and then a foreign language they were unfamiliar with, Mandarin. All the voices were of young men without obvious accents. The results showed that people with dyslexia were wrong 50 per cent of the time, whether the language was English or Chinese. People without dyslexia matched voice to avatars almost 70 per cent of the time when the language was English and 50 per cent of the time when the language was Chinese. The study shows dyslexia's strong link to phonological impairment and the interconnectedness of the brain processes involved in reading. But, Gabrieli consoles people with dyslexia, phonological impairment didn't present a great problem in the daily lives.

Sally Shaywitz, author of the widely acclaimed book *Overcoming Dyslexia: A New and Complete Science-based Program for Reading Problems at Any Level*, says that the study demonstrates the centrality of spoken language in dyslexia: that it's not a problem in meaning, but in getting to the sounds of speech.

According to Shaywitz, reading involves 'mapping' the images of written letters onto sounds that the reader already knows. People with dyslexia have trouble pulling apart the words they hear. If someone doesn't hear a spoken word as a group of component sounds, then the written word won't make sense. The word 'cat', for example, has only one syllable, but it has three units of sounds called phonemes: kuh, aah, tuh. Because people with dyslexia cannot decode words, they have difficulty in accessing information they have stored relating to that word. Shaywitz provides an example from real life: a child watching a baseball match says, 'I'm thirsty. Can we go to the confession stand?'

Brain-imaging studies show that the brains of people with dyslexia are wired differently. This different mode of brain organisation cannot be prevented or cured and doesn't go away over time. However, Gabrieli and Shaywitz both believe early intervention can help correct anomalies in brain activity in children with dyslexia.

In a study Gabrieli imaged the brains of children with and without dyslexia while they decided whether pairs of words rhymed. Two-and-a-half years later, he re-examined the reading ability of children with dyslexia. The children who had improved most showed most activity in the right prefrontal cortex, the area of the brain involved in visual memory, and the strongest connections between neurons in the white matter, which contains nerve fibres, in the same area. Skilled readers use the right prefrontal cortex less and less as they move from analysing words to reacting, almost instantaneously, to the whole word pattern by translating letters into sounds. This task requires language-processing areas of the brain, around and above the left ear.

This finding suggests that children with dyslexia who overcome their reading difficulties somehow bypass brain regions normally involved in reading. 'It seems they are better off using a completely different strategy,' Gabrieli says. He suggests that one possibility to help children with dyslexia would be to emphasise a more visual approach, similar to 'speed reading', as opposed to teaching them to translate letters into sounds. Other research indicates early practise in breaking words into phonemes, not simply reading words, helps children with dyslexia read better.

For the early detection of dyslexia, Gabrieli advises asking children to pick words that rhyme or begin with the same sound. Shaywitz also supports this approach. The inability to appreciate rhymes is an early indicator of dyslexia. 'If a child doesn't seem to "get" the funny rhymes in a Dr Seuss book, this may be the first sign of a reading disorder,' she says.

Boost your brain by
learning a second language

20
Bilingual bliss

The ancient Greeks regarded other languages as gibberish; ironically, nowadays when we say 'it's all Greek to me', we mean we do not understand it at all. 'Greek' meaning 'unintelligible language or gibberish' comes from Shakespeare's play *Julius Caesar*, and it reflects the deep-seated attachment that many of us feel towards our mother tongue. The noted American linguist Uriel Weinrich called it 'language loyalty'.

It may or may not be language loyalty, but until recently it was believed that teaching children a second language too early may impede 'normal' learning of their mother tongue. This belief is reflected in contemporary education practice which tends to offer formal schooling in a second language in later school years, not in the developmentally crucial toddler years of learning.

Neuroscience has now completely rejected the old hypothesis that the brain is set for one language only: learning a second language not only boosts children's brains during infancy, it also protects against decline in brainpower in old people.

Brain imaging of monolinguals and bilinguals shows that both process their individual languages in a fundamentally similar way: monolingual people and bilinguals show increased activity in language processing areas of the brain (left inferior frontal cortex). The one fascinating difference is that bilinguals appear to recruit more of the neurons available for language processing than monolinguals. This provides a fascinating insight into the language processing potential not used in monolingual brains.

Laura-Ann Petitto, a Canadian neuroscientist who is a leading researcher in the new discipline of educational neuroscience, says even bilingual parents often opt to 'hold back' one of the family's two languages in their child's early life. 'They believe that it may be better to establish one language firmly before exposing their child to the family's other language so as to avoid confusing the child,' she says. They also worry that earlier bilingual exposure may put their child 'in danger of never being as competent in either of two languages as monolingual children are in one.' Petitto's first-time use of the powerful functional near infrared spectroscopy brain-imaging technique (an imaging technique more suitable for studying young children than fMRI) to look into the developing brains of bilingual as compared with monolingual children has rejected unequivocally the myth that exposure to two or more languages 'too early' can cause developmental language delay and confusion. Her research also rejects the flip-side of this myth: later exposure is better.

Hungarian psychologist Agnes Melinda Kovacs and her Spanish colleague Jacques Mehler have opened up another fascinating window to the bilingual brain by studying 'crib bilinguals': young bilinguals are more flexible learners. Although infants in bilingual households have to learn roughly twice as much about language as their monolingual peers, the speed of learning is nearly the same for both. It seems that, far from being confused, infants in bilingual households develop superior mental skills or 'executive function' which play a critical part in complex social behavior. Executive function is our ability to control our thoughts and actions in order to respond suitably to our environment, as opposed to other brain functions dedicated to single tasks such as moving a finger or processing a sound.

Jared Diamond, an American scientist best known for his popular science books, agrees that the clearest difference identified by recent studies involves an advantage that bilinguals have over monolinguals, rather than disadvantage. 'Monolingual people have a special challenge involving executive function,' he says. 'Monolinguals hearing a word need only compare it with their single stock of arbitrary phoneme [sound] and meaning rules, and when uttering a word they draw from that single stock. But multilinguals must keep several stocks separate.'

Obviously, multilinguals have constant unconscious practice in using the executive function system of the brain. This unconscious practice gives older bilingual adults the ability to develop new strategies to process language, which helps compensate for age-related decline in cognitive power. A Canadian study has found that bilingual people tend to be diagnosed with Alzheimer's disease, the most common form of dementia, more than four years later than those who are fluent in a single language. The researchers believe that switching between languages strengthens the brain's 'cognitive reserve' — it can be compared with a reserve in a car tank which keeps you going a little longer when you run out of fuel. The idea of 'cognitive reserve' may also explain another finding: that older people who speak more than two languages are three times less likely to have memory problems than people who are bilingual. Older people are encouraged to start new brain-challenging activities such as playing bridge or solving Sudoku puzzles. These activities can engage your brain only for a few hours a day, while the bilingual brain is always engaged as it tries to limit interference from the other language to ensure the continued dominance of the intended language.

Well, more than half the world is bilingual. If you're not part of this world, it's time you joined the club.

Procrastination, the intentional delay of due tasks, is all too familiar to most of us

21

Putting off until tomorrow

Victor Hugo, who is best known for *Les Misérables* and *The Hunchback of Notre-Dame*, practised a novel way of dealing with procrastination: he would ask his valet to hide his clothes so that he would be unable to go outside when he was supposed to be writing. You don't have to work in the nude to force yourself to do that important task today, if you know the $E=mc^2$ equivalent of procrastination.

University of Calgary industrial psychologist Piers Steel is arguably the world's foremost expert on the subject of procrastination. He defines procrastination (from a Latin word meaning 'to put off until tomorrow') as willingly deferring something even though you expect the delay to make you worse off. After ten years' research (obviously without procrastination), Steel has come up with an equation to measure your desire to complete a given activity: what he calls utility or $U = EV/ID$, where E is the expectation of success, V the value of completing the activity, D the delay until reward, and I the personal sensitivity to delay.

If you enjoy an activity or value it, you are more likely to do it. The activity would have a high U because of high E (expectancy) and V (value). If the reward lies in the future (high D) or the person is impulsive or lacks self-control (high I), U shrinks. The author's high U in writing this book was influenced by high E (confidence in completing the book) and V (prospect of publication of the book), but low D (the looming deadline) and I (desire to finish the book).

Steel says procrastination is extremely prevalent, but 95 per cent of procrastinators wish to reduce it because they consider it as being bad, harmful and foolish. He urges that continued research into procrastination should not be delayed, especially because its prevalence appears to be growing.

Don Marquis, a celebrated New York humourist and newspaper columnist in the 1960s, has said: 'Procrastination is the art of keeping up with yesterday.' What happens in our brains when we follow this art? In ground-breaking research Laura Rabin and her colleagues at the City University of New York have linked the brain's prefrontal cortex to procrastination. The limbic system in the midbrain controls automatic functions — for example, it tells you to pull your hand away from a flame. The prefrontal cortex right behind the forehead is known as the 'executive' region of the brain. It integrates information and allows us to makes decisions. It's not automatic: you must kick it into gear ('I have to finish reading this page of the book'; once you're not consciously engaged in reading, the limbic system takes over). The researchers suggest that procrastination is a form of self-regulation failure because it has become automatic and ingrained.

Timothy A. Pychyl of Carleton University in Ottawa, Canada, says that procrastination is a dance between the brain and the situation. If you give in to what feels good, you procrastinate. 'Procrastination happens because you're disorganised, not very dutiful and probably impulsive,' he says. To increase the brain's executive function and decrease procrastination, he advises focusing on the problem of 'giving in to feel good' by first developing an awareness of this process and its negative effect on achievement. He also suggests using small goals that build on one another with regular deadlines and feedback (shorter deadlines would reduce D, the delay until reward, in Steel's formula). It's easy to procrastinate when goals are large and the path to them long and fuzzy, he says.

If it all fails, heed Mark Twain's advice: Never put off until tomorrow what you can do the day after tomorrow.

Motivation provides the will to win —
and dopamine helps in motivating the brain

22

All in a day's work

Dingy. Damp. Drafty. The three words that would have come to your mind if you were in 'le hangar' on a wintery Paris morning in 1898. Le hangar was a ramshackle wooden shed in a derelict backyard on rue Lhomond in the Latin Quarter. Once used as a morgue by students from a nearby medical school to dissect cadavers, the shed had a dirt floor covered with a scattering of asphalt, ruined brick walls, ill-fitting windows and a patched glass roof. If it rained, the roof leaked. Drops hitting the floor and the worn kitchen tables made soft but nerve-racking noise. A cast-iron stove with a rusty pipe delivered so little heat that in the depths of winter the temperature didn't rise above a chilly six degrees Celsius (43 degrees Fahrenheit). Chemical apparatus on tables and a blackboard on the wall were reminders that le hangar was a laboratory.

A slender 31-year-old woman was wearing an overcoat over her acid-stained smock to keep warm. The work she was doing was back-breaking. Le hanger was filled with large sacks, each containing as much as 20 kilograms (44 pounds) of pitchblende ore. The contents of each sack was ground, dissolved, filtered, precipitated, collected, re-dissolved, crystallised, re-crystallised. Because the shed had no chimney to extract noxious fumes, this work had to be done in the courtyard outside. In her own words: 'It was exhausting work to carry the containers, to pour off the liquids and to stir for hours at a time, with an iron bar, the boiling material in the cast-iron basin.'

A day's work for the love of science.

◆

The tall muscular man in his early 50s had come to Athens from his native Assos, a coastal town in Turkey, where he had spent his youth as a professional boxer. He no longer wanted to wrestle with men but with ideas. With only four drachmas in his meagre bundle of belongings, he had to find some work immediately. But he wanted to spend his daylight hours in the academy of Zeno, the Stoic.

The only night work he could find was to fill pails of water from a public fountain and deliver them to houses in the narrow, winding streets of Athens, most of which in 350 BC lay to the north-east of the Acropolis. Water carrying was done either by slaves or destitute housewives — in broad daylight. He found water-carrying work in the gardens of rich Athenians where he toiled late into the night. Torches were expensive and he had to work in the dark. The work enabled him to sit at the feet of Zeno. He continued working as a water-carrier even when he became well-known to Athenians for his simple life but high moral quality.

A day's work for the love of philosophy.

◆

These vignettes of the lives of Marie Curie, the discoverer of radioactivity and the first person to win two Nobel prizes, and Cleanthes the Water-Carrier, the Stoic philosopher who was the head of Zeno's academy for 32 years after Zeno's death, show that they were highly motivated individuals and loved something enough to make sacrifices for it.

Motivation is the force that drives us to achieve our goals; it gives purpose and direction to our actions. Curie's and Cleanthes's motivation didn't depend on rewards such as money, prestige or power (extrinsic motivation). It came from their desire to perform a task for the enjoyment it provided (intrinsic motivation). Rewards and punishment often backfire because they undermine intrinsic motivation. In an experiment — now a classic, which has been replicated many times with numerous variations — college students were either paid or not paid to work for a certain time on an interesting puzzle. Unpaid students played with the puzzle significantly more in a later unpaid 'free-time' than paid students, and also showed a greater interest in the task.

What happens in the brain when you are motivated to do a task?

First, a little dose of brain science (and it's not proverbial brain science). Four structures of the midbrain form what is known as the limbic system, which deals with urges and appetites. These structures are amygdala, hippocampus (important for forming new memories), thalamus (a kind of sensory relay station) and hypothalamus (regulates release of hormones). A network of neurons in these four structures — called the limbic loop — drives our decisions about whether or not to act on external and internal stimuli. A small pathway from the limbic system pumps the neurotransmitter dopamine into the frontal cortex, the 'executive' region of the brain right behind the forehead. When dopamine reaches the frontal cortex we feel good. Dopamine neurons play a vital role in brain networks that govern motivation and a sense of reward and pleasure; they are also associated with motor functions. These neurons go awry in Parkinson's disease, schizophrenia and drug addiction.

Recent studies suggest that dopamine is less about reward and pleasure than about drive and motivation. Common sense says that dopamine would be released when we perform a task and receive a reward. Experiments on baboons trained to perform a task and receive a reward show that dopamine is released just before the baboons perform the task and just before they receive the reward, not after. Even when the reward was not a sure thing but only a possibility, the release of dopamine increased substantially. However, when the reward was entirely expected on the basis of the preceding cue, dopamine neurons didn't respond to reward. Strangely, pleasure declines when we anticipate a reward.

Dopamine neurons in the questing minds of Curie and Cleanthes were firing not because of expectations of rewards but because they were intrinsically motivated to see new possibilities for the days ahead. Both had set goals they wanted to achieve but were open-minded about the future. Curie was trying to find a new element — radium — in the mountain of pitchblende ore she was refining. Cleanthes, like Zeno, wanted to achieve inner peace by being moderate in everything, the Stoic way.

The emerging view is that cultivating open-mindedness about the future leads to positive motivation. Dopamine, the molecules of motivation, excels at its task when you set a goal. But setting your mind on a goal and priming it with 'I will' may turn off the tap of your reservoir of intrinsic motivation (desire to perform a task for the enjoyment), especially if you think you would feel guilty or ashamed if you failed.

Your brain is tuned to
learning from your mistakes

23

You can become smarter, if you think you can

The mistakes of a learned man are a shipwreck which wrecks many others as it goes down, so says an old Arabian proverb. Everyone makes mistakes. Some learn from them. Others miss the opportunities to learn from their failures.

Now researchers have found that people who think they can learn from their mistakes have a different brain reaction to mistakes than people who think intelligence is fixed. Michigan State University researchers recruited undergraduates for their project. Each participant was wired to an EEG (electroencephalograph) to record electrical activity in the brain and given a task in which it was easy to make a mistake. They were asked to identify the middle letter of five letters series like 'MMMMM' or 'NNMNN' which repeatedly flashed on a computer screen. Sometimes the middle letter was the same as the other four, and sometimes it was different. In either case, the participants were asked to push a button. They had only a few milliseconds to make a decision.

When you make a mistake, your brain makes two distinct reactions which can be recorded on EEG. An initial reaction appears about 50 milliseconds after the mistake, and the second arrives anywhere between 100 to 500 milliseconds after the mistake. The first reaction is mostly involuntary and the second indicates that the person is consciously aware of the mistake and trying to learn from it.

The results of the experiment showed that the participants who thought they could learn from their mistakes did significantly better after making mistakes. Their EEG signals showed a bigger second signal, the one that implied, 'I see that I've made a mistake, so I should pay more attention to it.'

'This finding is exciting in that it suggests people who think they can learn from mistakes have brains that are more tuned to pick up on mistakes very quickly,' says Jason Moser, the lead researcher. Their brains are tuned differently at a very fundamental level. What you need is the right 'mind-set': this mind-set is not simply dependent on intelligence; motivation and effort are larger part of it. On the other hand, people who think that they can't get smarter will not take opportunities to learn from their mistakes.

The researchers believe that these findings could help in training people to believe that they can work harder and learn more, by showing how their brain is reacting to mistakes. All they need is EEG caps on their heads.

In her famous 2008 Harvard Commencement address, J.K. Rowling, author of the mega-selling *Harry Potter* books, talked to students about the fear of failure: 'You might never fail on the scale I did, but some failure in life is inevitable. It is impossible to live without failing at something, unless you live so cautiously that you might as well not have lived at all — in which case, you fail by default.'

If you learn from a failure, it's no longer a failure. You have learned something new. You have become smarter. The eminent Danish physicist Niels Bohr summed up this essential lesson of learning when he said, 'An expert is a person who has made all the mistakes that can be made in a very narrow field.'

Praise children for qualities
they can control, like effort

24
Wow, you're a little Einstein

If you're one of those parents who love praising their children for their intelligence, believing that the flattering words would enhance their self-esteem and performance, you're wrong. More than three decades of research shows that calling your children smart can undermine their ability to learn. 'Focus on effort — not on intelligence or ability — is the key to success in school and life,' says American research psychologist Carol Dweck who has spent decades studying how people cope with failure.

In 1998 Dweck was the lead author (with Claudia Mueller) of a now-famous study which compared the motivation and performance of children praised for intelligence with those of children praised for making an effort. In the study, six groups of children aged 10 and 11 were given a series of tasks with different problems to solve. After the first task, they all were told they did very well, no matter how good was their score. Some were told, 'You must be smart at these problems,' while others were told, 'You must have worked hard on these problems.' Subsequent tasks became harder, and children were later allowed to choose tasks. The results showed striking differences between the two groups. Children praised for their intelligence worried more about failure than children praised for their effort. When they didn't do so well, 'smart' ones blamed a lack of intelligence, whereas those praised for working hard blamed a lack of effort. When allowed to choose tasks, 'smart' ones chose tasks they knew they would do well on; 'hard workers' chose tasks they thought might help them learn something new. The 'smart' ones also preferred to find out about the performance of others on the tasks rather than learn new ways of solving problems.

When Dweck and Muller designed their experiment, they expected that different forms of praise — 'smart' or 'hardworking' — would have a rather modest effect. They were surprised by the magnitude of difference between the two groups. 'Emphasising effort gives a child a variable that they can control,' Dweck explains. 'They come to see themselves as in control of their success. Emphasising natural intelligence takes it out of the child's control, and it provides no good recipe for responding to failure.'

The study's message, in essence, is praise children but for work and persistence, not intelligence. Even geniuses work hard, very hard. They succeed because of their persistence, not intelligence alone. Heed Edison's words — genius is 1 per cent inspiration and 99 per cent perspiration — when praising children.

Finally, the children praised for intelligence described it as a fixed trait more than children praised for hard work, who believed it to be something that is malleable and can be developed over time. Dweck's recent research shows that students' mind-set about their intelligence — whether it is a fixed or a changeable quality — guide their motivation and learning. 'Students who hold a fixed view of their intelligence care so much about looking smart that they act dumb,' Dweck says, 'for what could be dumber than giving up a chance to learn something that is essential for your own success.'

The word 'belief' is the clue here. If you believe you can develop your intelligence over time (a growth mind-set, according to Dweck), you can be as intelligent as you want to be. If you have believe you were born with a certain amount of intelligence (a fixed mind-set), that's the end of the road for you.

IQ — intelligence quotient — fails to measure all aspects of real-world intelligence

25

Wish to raise your IQ? But first ...

Intelligence is difficult to define. Dolphins (capable of abstract communication), primates (can use simple tools) and African grey parrots (can categorise objects), for example, are intelligent in their own ways, but they all lack the most important aspect of human intelligence: creativity which has resulted in the development of technology.

If technology defines the intelligence of humans as a species, what defines the intelligence of individual humans? Or, what does it really mean to be intelligent? Eric R. Kandel, the winner of the 2000 Nobel Prize in Physiology or Medicine, pinpoints it 'to be able to think deeply about problems, to be able to analyse new problems, to see relationships between events, to be creative.' There's no absolute measure of these characteristics of intelligence. The most famous — not necessarily the perfect — measure are IQ tests. As real-world intelligence has many dimensions, the tests fail to measure all of them. We can be more intelligent in some things and not in others. No one is equally intelligent in everything.

Sometimes even highly intelligent people do foolish things. Keith E. Stanovich, author of *What Intelligence Tests Miss: The Psychology of Rational Thought*, explains this phenomenon as 'the inability to think and behave rationally despite having adequate intelligence'. He calls it 'dysrationalia' (analogous to 'dyslexia') and attributes two causes to it: (1) people tend to be cognitive misers — that is, they take the easy way out when trying to solve a problem; and (2) mind-ware gap, which occurs when we lack the specific knowledge, rules and strategies to think rationally. IQ tests fail to measure dysrationalia; however, you can test your cognitive miserliness by taking this simple test devised by Nobel Prize-winning psychologist Daniel Kahneman and his colleague Shane Fredrick: 'A bat and ball cost $1.10 in total. The bat costs $1 more than the ball. How much does the ball cost?' If you are a cognitive miser, a good IQ score is no guarantee that your answer would be correct (the correct answer appears at the end of the story).

Intelligence is not related to the size of the brain or the numbers of neurons in the brain. There is no single intelligence centre in the brain; intelligence is built by a network of regions across both sides of the brain. Intelligence is how fast neurons are making connections, how well information is travelling throughout the brain. Richard Haier, an American psychologist who has been using brain imaging to discover the neural basis of intelligence, says intelligence is characterised by individual differences in learning, memory and attention and how they are integrated in any one individual, and the brain can generate the same IQ scores a number of ways. He believes that one day we may be able to estimate someone's IQ and other intelligence factors from a brain scan.

Is there a difference between lower and higher IQ brains? Brain scans do show a difference. The lower IQ brains show lots of activity: they try harder. The higher IQ brains show less activity: they try smarter, not harder. 'It's easy for them, relatively speaking,' says American cognitive neuroscientist John Gabrieli. 'Smarter brains, simply put, are more efficient.'

From biographical profiles researchers have estimated the IQ of forty-two past US presidents (from George Washington to Bill Clinton). They all scored at least 130 (which puts them in the top 2.2 per cent of the population). If your IQ score is below this range or any other range (the average score is 100; genius range starts from around 135), heed the findings of a growing number of studies: mental and physical exercises can help raise your IQ score by a few points. Let's start with the evidence of the benefits of physical exercise ...

Answer: You're a cognitive miser if you answered 10 cents. In fact, the bat costs $1.05 and the ball 5 cents.

Follow the old Latin saying 'mens sana in corpore sano' (a healthy mind in a healthy body) to increase your brainpower

26

Exercise your body — and build your brain

If you focus on physical exercise to trim the waistline or lose weight or keep you body fit, there is good news for you: it's also making you smarter by boosting your mood, memory and learning.

How might exercise help the brain? Exercise pumps more blood to the brain, which gives more oxygen to the neurons and thus making them better nourished. Exercise also causes the release of proteins known as growth factors, including one called brain-derived neurotrophic factor (BDNF). As the levels of BDNF build up neurons start to branch out and build new connections in the hippocampus, the brain region important for memory. These new connections signify a new fact or skill that has been learned and stored for future use. As we age, individual neurons start to die. We know now that this loss is not permanent: the brain can make new neurons. Again, BDFN plays a role in growing new neurons — and exercise helps in building up its levels by increasing blood volumes.

There is another way to look at the benefits of exercise. A plethora of studies have shown that physical activity reduces the risk of heart disease, type 2 diabetes, colon and breast cancer, depression and osteoporosis. At the same time these diseases have shown to compromise mental abilities. 'Therefore, you might expect that increased physical activity and exercise would maintain cognition by reducing the risk of diseases associated with cognitive decline,' write American psychologist Christopher Hertzog and others in *Scientific American Mind*.

Exercise is a highly effective way to reduce the risk of dementia later in life. A

University of Illinois study of a group of 55- to 80-year-olds who exercised regularly for a year — 40 minutes' walk three times a week — showed about 2 per cent increase in the volume of specific regions of the hippocampus. Another group of 55- to 80-year-olds who did a toning workout, which included weight training, yoga sessions and stretching, for 40 minutes three times a week for a year lost about 1 per cent of the volume of the same region of the hippocampus. The study shows that aerobic exercise — not toning exercise — is good for hippocampus, which is one of the first brain regions to start showing decay in Alzheimer's disease. 'Although it used be thought that ageing was a one-way street that was going the wrong direction, we know from our work and other work that that's not the case,' says Arthur Kramer, co-researcher of the study. Other studies have shown that people who exercise regularly in middle age are one-third less likely to get Alzheimer's disease in their 70s as those who did not exercise. Even people who begin to exercise in their 60s reduce the risk by half.

John Ratey, a Harvard Medical School psychiatrist and author of *Spark: The Revolutionary New Science of Exercise and the Brain*, thinks of exercise as medication for AD/HD (attention deficit/hyperactivity disorder) For some it may actually be a replacement for stimulants, but 'for most, it's complementary — something they should absolutely do, along with taking meds, to help increase attention and mood,' he says. For children with AD/HD, he advises, team activities or exercise with social component are especially beneficial.

Obviously, the brain is like a muscle and we need to exercise it, but for how long and how often? The advice used to be 30 minutes at least five days a week, but experts now say aim for 150 minutes per week and divide it into whatever portions (of at least 10 minutes' duration) you like. You don't have to drag yourself to the gym to do strenuous exercise: even walking has shown the same effects. You may juggle if you prefer (but its benefits are different). An Oxford University study has found that practising a task such as juggling can increase the brain's white matter, the nerve fibres that connect different parts of the brain. Changes in grey matter, which consists of tightly packed neurons, follow new experiences and learning. This study is the first to show changes in white matter.

Studies on mice show that the benefits of exercise last only a few weeks after exercise has ended. To stay sharp, you have to exercise regularly. Don't lose sight of that exercise bike.

Computerise or not to
computerise mental workouts

27
Brain training: hope or hype?

During the past two decades the idea of neuroplasticity, that our brains are capable of rewiring even late in life, has introduced a new industry of brain-training or brain-fitness computer programs to improve memory and keep the mind sharp. The multimillion-industry industry's hypothesis is 'use it or lose it': the more you use your brain, the more powerful it will become. Many scientific studies back up this hypothesis. The big question is whether it is worth investing in brain-training programs to exercise your brain or it is better to exercise your brain on your own (for example, by learning a new language or learning to play bridge, chess or a musical instrument).

The largest study of brain-training programs so far — it involved 11,430 volunteers aged from 18 to 60 — suggests that there is no scientific evidence such programs sharpen mental powers. The volunteers were divided into three groups, and each group practised a series of online tasks for a minimum of ten minutes a day, three times a week, for six weeks. The first group worked on tasks that emphasised reasoning, planning and problem-solving activities. The second group focused on short-term memory, attention, visuospatial processing and mathematics tasks commonly found in commercial brain-training programs. The third group, the control group, simply used the internet to find answers to a set of obscure questions.

The research team — led by Adrian Owen of the UK's Medical Research Council Cognition and Brain Services Unit — found that although all the volunteers showed improvements in the tasks they were assigned, there was absolutely no difference between the groups on other tests of mental abilities. Owen concedes that the study findings do not necessarily mean that brain-training programs are pointless, but the evidence in their favour is not strong. However, an earlier study funded by Posit Science, a brain-training software company co-founded by Michael Merzenich, a neuroscientist at the University of California, showed modest effect. No large-scale study has shown that brain-training programs are the fastest way to boost mental skills.

Brain-training needs to be put under more tests before they can be proven science or nonsense. In the meantime, learn new words to increase grey matter in your brain. That's the conclusion of a study conducted by a team of Chinese neuroscientists and linguists, led by neuroscientist Veronica Kwok of University of Hong Kong. Nineteen Beijing University students spent about two hours spread over three days learning new words to describe colours. They saw four colours (two shades of green and two shades of blue) on a computer screen at 1-second intervals, and heard corresponding made-up Chinese monosyllables (áng, sòng, duān, kěn) assigned to these colours. Their task was to judge, by choosing 'yes' or 'no' on a printed form, whether the sounds they heard matched the new names for displayed colours. By the end of the three days of lessons, students could accurately describe each of the colours by their made-up names. Students' brain scans before and after the tasks showed detectable increases in the volume and density of grey matter. The finding shows that the grey matter in the adult brain can change very quickly, specifically during learning new words.

Forget the brain gym, learn new colour names. Go to www.en.wikipedia.org/wiki/List_of_colors for a comprehensive list of colours, including colour swatches. You can design your own brain-training program. The good thing about brain training is that it does no harm, whether it's free or comes in an expensive package.

28

The choice is yours

Some wit has said: it's choice, not chance, that determines your future. As I write these words, it's not the future I'm worried about, but choosing ideas on decision-making I must cover in a few hundred words in this book. The prefrontal cortex in my brain, the region responsible for decision-making, is now active. The choice is not simple. The amount of research on decision-making is mind boggling. Information overload caused by mind-numbing choices has resulted in information fatigue. I shuffle through research papers on my desk and gather more information; the activity in the prefrontal cortex increases. I open another bulging folder; my mind experiences cognitive and information overload. My frustration and anxiety skyrocket. If I were lying inside an fMRI machine, the brain scan would now show a sharp decline in the activity in the prefrontal cortex as if the fuse had melted. The prefrontal cortex is no longer playing its other role of keeping emotions under check. My brain's emotional regions have gone hog wild. Too much information is not helping me to make a well-informed and balanced decision.

Forget the witty quote, let me try chance. Toss a coin? No good. The law of averages says that in the long run heads and tails even out. What about placing my decision-making in the lap of gods? Luck is not the same as chance. Being lucky is an easy skill to learn, at least according to Richard Wiseman, a UK psychologist. In his book, The Luck Factor, he describes four strategies for creating good fortunes (supposing luck can be tied to the right decisions):

- Decisions informed by intuition (*next story*) are likely to produce happier outcomes: you can boost your intuitive abilities by meditating to clear your mind of other thoughts.
- Don't be a creature of routine. New experiences offer new opportunities.
- Be positive. Expect good fortune. These expectations become self-fulfilling prophecies.
- Turn bad luck to good. Do not dwell on ill fortune. Take control of the situation.

Wouldn't it be a good idea, if as a science writer I place my decision-making in the lap of science, instead in the lap of gods? Think like a scientist: make a hypothesis; collect information; reconsider your hypothesis in the light of information. Sounds boring? Try PROACT (problem, objectives, alternatives, consequences, trade-offs), a system developed by psychologists:

- Define your problems.
- Specify your objectives.
- Consider alternative course of actions.
- Evaluate the consequences of all possible decisions.
- Look at the trade-offs of each course of action.

Ockham's razor — a rule developed by William of Ockham who lived in the 13th century in England — can also help in thinking clearly. The rule — it is vain to do with more what can be done with less — implies keep the number of causes and explanations in your decision-making to a minimum. Sometimes it's better not to devour too much information. Too many choices can make you miserable.

American neurobiologist Antonio Damasio considers emotions integral to our decision-making processes. If we don't have those gut responses, we'd be caught in an endless cycle of analysis, drawing infinite pros-and-cons lists in our heads, he says. Damasio's hypothesis is built around somatic markers: the brain 'tags' information involved in decision-making with markers based on emotions. The markers are bodily responses, such as your gut tightening in fear or your shoulders convulsing with warm laughter. They are used consciously or unconsciously, to guide our decision-making. 'It's not that I'm saying the emotions decide things for you,' he says. 'It's that the emotions help you concentrate on the right decision.' Although emotions may give you head start, it's best to avoid making important decisions when you are emotional.

It's also best to avoid making decisions when you are hungry. Some studies show that low glucose levels (our bodies derive energy from glucose which is manufactured from all kinds of food) will lower willpower. Experiments show there is a finite store of mental energy for exerting self-control. Therefore, low glucose levels are likely to affect our self-control. 'Even the wisest people won't make good choices when they're not rested and their glucose is low,' warns psychologist Roy F. Baumeister, co-author of *Willpower: Rediscovering the Greatest Human Strength*. More than 2500 years ago, Gautama Buddha, the Enlightened One, refused to continue his sermon when he discovered one of his followers hadn't had his meal, saying 'he wouldn't be able to follow me.' So, the ages-old wisdom — now supported by new research — is: don't make an important decision on an empty stomach.

No decision is foolproof: there's always a risk of making a wrong choice. Whether you make a wrong decision or a right one, your brain makes a decision ten seconds before you realise that you have made it. Researchers have arrived at this conclusion after looking at brain activity of volunteers while they performed a decision-making task. This finding challenges the 'consciousness' of our decisions.

Was my decision to write these words the right one? Research has shown that younger adults are better decision-makers than older ones. I fall into the latter group, so I made a wrong decision. No. Now researchers at Texas A&M University say that 'older adults are better at evaluating the immediate and delayed benefits of each option they choose from' and are better at creating strategies in response to the environment. Earlier studies tested the ability to make decisions one at a time without regard to past or future. The Texas A&M researchers' model was more like decision-making in the real world and was tested on two groups of younger (college adults) and older (ages 60 to early 80s) people.

You can make you own real-world decision on whether to continue reading this text about decision-making was the right decision.

Intuition — thoughts and feelings that come to mind immediately and without reasoning — can help as well as harm us

29

That gut feeling

Intuition and reasoning are two different modes of thought. Intuition is fast, automatic and effortless. Reasoning is slow, controlled and requires effort to employ.

Intuition is our unconscious mind speaking to us. 'The mind gathers information that is not conscious and then finds a way to communicate it to us through "intuitive feelings",' says Malcolm Gladwell, author of Blink: *The Power of Thinking Without Thinking*.

We cannot immediately explain the reasons for these gut feelings and make automatic decisions because they 'feel right' at the time. Experienced chess players, for example, often know the right move to make even if they cannot pin point why. Similarly, you meet a charming stranger at a pub and you have a sneaking feeling that the person is not trustworthy. Your intuition might have turned out to be your best defence by keeping you out of harm's way if you later discovered that guy was indeed a swindler or worse.

Not always. 'Intuition is powerful, often wise, but sometimes perilous, and especially so when we overfeel and underthink,' warns David G. Myers, author of *Intuition: Its Powers and Perils*. This overfeeling and underthinking may lead to intuitive prejudices (*see page 97 to find out how hidden biases can colour our decision-making*). Myers provides a quick quiz to illustrate that even smart people make predictable and sometimes costly intuitive errors: 'In English words, does the letter k appear more often as the first or third letter?' Answer it before you read Myers' explanation below:

For most people, words beginning with k are more immediately available in memory. Thus, using the 'available heuristic', they assume that k occurs more frequently in the first position. Actually, k appears two to three times more often in the third position.

Researchers at the UK's Medical Research Council Cognition and Brain Services Unit have found that the trustworthiness of our intuition is really influenced by what is happening physically in our bodies, and the ability to make intuitive decisions varies considerably with different people. The researchers asked volunteers — wearing heart and sweat monitors — to try to learn how to win a card game they had never played before. They game had rules but no obvious winning strategy. The players had to follow their gut feelings. The researchers tracked the players' success over time as well as changes in their heart rate and sweating. The players also reported whether they relied on intuition or reason. The researchers found that those who had become most successful the most quickly also turned out to be the ones who had higher heart and perspiration rates. 'What happens in our bodies really does appear to influence what goes in our minds,' comments Barnaby D. Dunn, the lead researcher. 'We should be careful about following these gut instincts, however, as sometimes they help and sometimes they hinder out decision-making.' Simply put, the mind and body are inextricably linked.

In fact, intuition and reasoning are complementary, not competitive. 'Without intellect, our intuition may drive us unchecked into emotional chaos,' writes Michael Shermer, in his 'Skeptic' column in *Scientific American* magazine. 'Without intuition, we risk failing to resolve complex social dynamics and moral dilemmas.'

Also, researchers are finding that the notion that women are really more intuitive than men might actually be a myth. And the fact: instincts of animals are much sharper than humans. Trust you dog, if you're worried about trusting your gut feelings.

30

The clue words are 'brain' and 'thunder'

Ask the group to think of a word that follows each of the two words 'brain' and 'thunder' to turn them into compound words? In fact, you don't need a group brainstorming session to find the word. Even if you do, business psychologist Peter Heslin of Southern Methodist University in the US would tell you that brainstorming sessions are ineffective. We assume he wouldn't be surprised if the group comes up with 'less' for an answer instead of 'storm', because he knows brainstorming can choke creativity.

'Despite its immense popularity, when groups of people interact for the purpose of brainstorming, they significantly overestimate their productivity and produce fewer unique ideas than normal groups of people generating ideas alone,' he comments, and proposes an alternative to brainstorming — brainwriting.

The process of brainwriting is simple. Each group members writes ideas on a slip of paper in silence, using a different coloured pen, and passes it to the member on their right. The right-hand member reads the idea and adds his or her own idea. Once a slip has four or five ideas, it's placed in the centre of the table. When all slips are done, members analyse and discuss them. Most popular ones are recorded. The final stage involves members working alone for 15 minutes to generate yet more ideas.

Heslin's research shows brainwriting yields better ideas than brainstorming. When you are brainstorming, you have to wait a few minutes to allow others to talk before you can reveal you own ideas. 'People will lose confidence and start thinking their ideas are unworthy or crazy,' he says. He has found that groups that contain people with diverse but overlapping knowledge and skill tend to be particularly creative in brainwriting sessions.

Dutch psychologists Bernard A. Nijstad and Wolfgang Stroebe offer some practical recommendations for brainstorming:

- Keep groups with verbal idea-sharing small and, if necessary, split up larger groups.
- Discourage people to explain their ideas in detail.
- Electronic brainstorming and brainwriting are to be preferred above verbal brain storming.

The Individual and Interactive Mind

Self is always obvious to us,
but how does it emerges from the brain?

31

Looking for 'I' in the brain

Being self-less is considered a good thing, but too much of self makes us selfish. A realistic self-image relates to a healthy mind, but a defective self-image defines an anxious and self-doubting mind. Self-reflection helps us to keep our self-image in step with reality, but our individual abilities for self-reflection and self-insight vary enormously. What is the self?

Alvaro Pascual-Leone of the Harvard Medical School has a simple test to explain the meaning of self-awareness. When you dab an odourless dye on the eyebrow of a monkey trained to look in a mirror, the money will try to rub the dye off the mirror. Try the same with a chimpanzee; it will wipe its own eyebrow. The chimp is aware of its own mental state, the monkey is not. Pascual-Leone points out that the chimpanzee can access a whole series of social behaviours that lie outside the mental reach of the monkey.

Without the help of an ape making faces at a mirror, the ancient Greek philosophers speculated that the self determines behaviour. The way we identify with others and distinguish between self and others plays an important role in our social development. We know now that self-awareness is the product of the brain. 'The "I" is a construction — a story that a brain tells, a fantasy that it weaves — an illusion of a real self who has consciousness and free will,' says Susan Blackmore, a British writer. 'This fantasy is constructed all the time we are awake and our brain is functioning normally.'

Neuroscientists are now trying to find out how 'I' emerges from the brain. But first, the incredible story of Phineas Gage, which changed our understanding of the relation between mind and brain. Gage was an American railroad foreman and in 1848 an explosion drove a tamping iron (1.1 metres/3 feet 7 inches long and weighing 6 kilograms/13½ pounds) through his left cheek and out through the top of his head, landing many metres behind him. Although most of the front part of the left side of Gage's brain was destroyed, he didn't lose consciousness. A few minutes later he was sitting in an ox cart and writing in his workbook. He was treated in a hospital by John Harlow, a young doctor, with such a success that he returned to his home after ten weeks.

What gave Gage a permanent place in the annals of neurology was the change in his personality that Dr Harlow's subsequently observed: 'A child in his intellectual capacity and manifestations, he has the animal passions of a strong man ... His mind was radically changed, so decidedly that his friends and acquaintances said he was "no longer Gage".' After about 150 years, a computer-generated three-dimensional reconstruction of the skull of Gage showed that his prefrontal cortex was damaged. This region of the brain plays an important role in the ability to control our behaviour.

V.S. Ramachandran, one of the brightest minds in neuroscience today, suggests that groups of neurons called mirror neurons are critically involved in self-awareness. Mirror neurons are a type of neuron that respond equally when we do something and when we see someone else doing the same thing.

Mirror neurons were discovered by the Italian neuroscientist Giacomo Rizzolatti and his colleagues at the University of Parma when they recorded the brain activity of monkeys performing certain tasks. They found that individual neurons would only respond to very specific actions; for example, when pushing a button or pulling a lever. And they noticed something more surprising: when a monkey reached for a peanut a certain neuron would fire. Amazingly, when the monkey saw another monkey reach for the peanut, the same neuron fired again. 'It was doing a sort of internal virtual reality simulation of the other monkey's action in order to figure out what he was "up to",' comments Ramachandran, 'It was, in short, a "mind-reading" neuron.'

The actions of mirror neurons are involuntary and automatic. We don't have to think about what other people are doing or feeling, we just know it. Ramachandran suggests that self-awareness is simply using mirror neurons for 'looking at myself as if someone else is looking at me'. 'The mirror neuron mechanism – the same algorithm – that originally evolved to help us adopt another's point of view was turned inward to look at your own self,' he says.

It will be awhile before we know the final answer to the neural mechanism of self-awareness. But what is self-awareness good for? According to Uwe Herwig, a Swiss psychologist, it helps us recognise and manage fear, anger and other potentially destructive emotions. Recent brain-scanning experiments show that when people distance themselves from upsetting feelings, the rational parts of their brains (such as the prefrontal cortex) tamp down emotional ones (such as the amygdala) — and they feel better. 'As imaging technology continues to develop,' he hopes, 'it is possible that brain-scanning devices might someday provide real-time feedback to people as they meditate, enabling them to train their brain to be more mindful.' This kind of feedback might be used to help people to master emotional self-regulation.

32
Feeling for others

Empathy is an emotional reaction that makes us sensitive to the feelings and thoughts of others. The word was coined in 1909 by British psychologist Edward Titchener. Before that the word 'sympathy' included the current meanings of 'sympathy' as well 'empathy'. So what is the difference? 'For most of us today, empathy differs from sympathy in the way that "I feel your pain" (empathy) differs from "I feel sorry about you being in pain" (sympathy),' explains US philosopher Michael Slote.

Titchener used the word 'empathy' to mean our ability to relate to the experiences of other persons by mirroring them into our minds. Interestingly, some neuroscientists now explain empathy in terms of mirror neurons (*previous story*). You see a stranger stubbing her toe and you immediately shrink in pain. Mirror neurons in your head enable you to feel what's in her head. The brain is not using logical thought processes to interpret and predict her emotions, as previously believed. Instead, proponents of mirror neuron theory say, mirror neurons in your brain 'simulate' her emotions which trigger your empathic emotions. In people with autism the mirror-neuron system may be not fully functional, which may explain why they are unable to respond to feelings of others.

Simon Baron-Cohen of University of Cambridge has developed a scale to measure empathy. His Empathy Quotient (EQ) measurements show that in the general population EQ varies along the familiar bell-shaped curve showing normal distribution. The curve reveals that females score slightly higher on EQ than males. A child version of the EQ also shows that on average girls have a slightly higher EQ than boys. Why do people have different levels of empathy? 'The most immediate answer is that it likely depends on the functioning of a special circuit in the brain, what I call the "empathy circuit",' says Baron-Cohen. The so-called empathy circuit has ten interconnected brain regions. He believes that empathy involves some form of mirroring other people's actions and emotions. 'The mirror neuron system in humans is hard to measure, obviously because it is unethical to place electrodes into the awake human healthy brain,' he says.

There are indirect ways to study the role of mirror neurons in empathy. In an experiment Lisa Aziz-Zadeh of the University of Southern California observed brain scans of a healthy woman who was born without limbs as she watched videos of people performing actions such as holding and eating an apple slice, sewing with a needle and tapping a finger. The woman couldn't perform the tasks herself but her mirror neurons fired up and activated parts of the brain that control movement. 'What's interesting is that even when she can't do it, when it's impossible for her, she still recruits her mirror system, but she additionally recruits these mentalising regions,' Aziz-Zadeh says. 'Mentalising' is a process in which a person tries to understand what someone else is thinking. Aziz-Zadeh's study provides a clue that our brains work hard to understand and empathise with others — no matter how different they may be.

Empathy may be hardwired but a recent study shows that empathy is in decline among young people in the US. Sara Konrath of the University of Michigan says that college students today show 40 per cent less empathy than students in the 1980s and 1990s. The reason perhaps is that people are having fewer face-to-face interactions, instead communicating through social media such as Facebook and Twitter. 'Empathy is best activated when you can see another person's signal for help,' she says.

It may come as a surprise to you but US studies show that wealthy people show less empathy compared with those from lower socioeconomic classes. According to Dacher Keltner of the University of California, people from lower socioeconomic classes are threatened by the environment, by institutions and by other people. 'When you're more vulnerable, you solve problems by turning to others,' he says.

The most peculiar feature of yawning is its contagiousness; and contagious yawning reflects empathy (*page 140*).

33

Did you say 'I …'?

The personal pronoun 'I' is the most commonly used word in spoken English. Who uses it more?

1. Leaders or followers?
2. Rich or poor?
3. Men or women?
4. Truthtellers or liars?
5. Depressed or happy people?

The answers at the end of the story will surprise you. These questions have been adapted from a test devised by James W. Pennebaker, a US social psychologist internationally recognised for his computer analysis of texts for their psychological content. The 'texts' include books, speeches, letters, poems, blogs, emails, tweets, class writing assignments and conversations. He has even analysed his own writings. 'There is something creepy about analyzing your own emails, letters of recommendations, web pages, and natural conversations,' he admits.

In his book, *The Secret Life of Pronouns*, he argues how the smallest, most commonly used words such as 'I', 'she' and 'who' serve as windows into our thoughts, emotions and behaviours.

'You can be focused on yourself for many reasons,' he says. 'Once you appreciate that

"I" tracks attention, you see it's a powerful marker of a speaker's psychological state.' Some of Pennebaker's other observations (which have been paraphrased) on the use of 'I':

- Confident people don't talk about themselves: their usage of 'I' is less. Less confident people use hedging phrases ('I think ...').
- Higher status people use 'I' less than lower status people. Higher status people tend to be more comfortable with themselves and are less self-conscious than the more insecure lower status people.
- Most people assume that men use 'I' more than women. In fact, about 14.2 per cent of women's words are personal pronouns compared with 12.7 per cent for men. This is a huge statistical difference. Men and women use language differently because they negotiate worlds differently.
- Liars avoid 'I'. When you're lying, you almost distance yourself from the words. You're not owning your statements.
- People who are clinically depressed tend to use 'I' at very high rates — almost as though they are embracing their unhappiness. Once you snap out of your depression, your use of 'I' will drop.

Pennebaker is also a pioneer in the field of therapeutic writing (*page 51*).

Answers: 1. followers; 2. poor; 3. women; 4. truthtellers; 5. depressed people

The human brain seems to have been built for optimism — it sees the doughnut, not the hole

34

The glass is always full

Winston Churchill once said, 'For myself, I'm an optimist. It does not seem to be much use being anything else.' Now neuroscientists say that our brains, like Churchill's brain, have optimism bias. Optimism is essential to our survival — there's not much use being anything else — and the brain is hardwired for optimism.

Tali Sharot and her colleagues at New York University appear to have found neuron networks which are responsible for optimism. Using an fMRI scanner, the researchers recorded brain activity in volunteers when they thought about positive future events in their everyday life (such as winning a large sum of money). Two regions deep within the brain lit up: the amygdala, responsible for emotions, and parts of the anterior cingulate cortex, which boost the flow of positive emotions. The same two regions are less active in depressed people. If the brain didn't have bias towards optimism, we would be prone to increased anxiety and depression.

'We need to burst a giant bubble — the notion that we perceive the world as it really is,' Sharot writes in her book, *The Optimism Bias: A Tour of the Irrationally Positive*. Instead, we see the world through rose-tinted glasses. And we are born with them. If Sharot and other neuroscientists are right, we should not be seeing holes in doughnuts ('Between the optimist and the pessimist, the difference is droll. The optimist sees the doughnut; the pessimist the hole,' according to Oscar Wilde).

In another study Sharot tried to find out what was going in the brains of people who remain unrealistically optimistic even when every bit of news is gloomy. While lying in the fMRI scanner, volunteers were presented with a series of negative life events (such as being robbed, getting fired from their job, developing cancer). After each scenario, the volunteers were asked to guess the odds of the event happening to them in the future. They were then told the true average probability of the event while they were still in the scanner. In the second part of the study, participants were asked again to guess the odds for experiencing each scenario, but this time they were not in the scanner.

Sharot found that the volunteers updated their initial estimates based on the information given, but only if the true figures were better than expected (for example, if their estimate of being robbed was 30 per cent and the average probability was 20 per cent, they might reduce their estimate to, say, 22 per cent). If the estimate was worse than expected, then the participants largely ignored the true figures. The brain scanned showed increased activity in the frontal lobes of the brain, the region of the brain associated with emotional control, when the true figures were better than expected. This indicated active processing of the new information to recalculate estimate. But when the new figure was worse than expected, participants who had rated highest for 'optimism' on a personality questionnaire showed the least activity in the frontal lobes, indicating that participants were disregarding the new evidence. The study shows those who rated highest for optimism seemed to simply refuse to perceive risks. Optimism can be a positive thing, says Sharot, because 'it can lower stress and anxiety, and be good for health and wellbeing.' But she also sees the downside: we are less likely to take precautionary action, 'such as practising safe sex or saving up for retirement.'

Optimism is related to the placebo effect. It can also stimulate your immune system and therefore be beneficial for your health. However, some studies show that positive thoughts do not always lead to positive effects.

Overall, optimism is good for you, but there are limits to seeing the glass half full. Unrealistic optimism could lead you to ignore real dangers — not seeing that glass is really empty.

35

Small acts make you — and others — happy

'Small differences in the initial conditions produce very great ones in the final phenomena,' observed Jules-Henri Poincaré, the renowned French mathematician and philosopher of science, in 1908. The observation received little attention from his contemporaries, but has now earned him the title of the 'founder of chaos theory'. The first study of chaotic behaviour in nature was made by the American meteorologist Edward Lorenz in 1963, when he developed a computer model to predict weather patterns. He was surprised to find that even a small change in initial values resulted in wildly different conditions in his predictions. This is sometimes called the 'butterfly effect': an action as small as a butterfly flapping its wings, say in New York, could bring about a snowstorm weeks later on the other side of the Atlantic in London. Chaotic behaviour occurs in phenomena as diverse as the stock market, population changes and the human heartbeat.

There is evidence that people who care more about others are happier than more selfish people. Also, small acts of kindness do not only have an enormous effect on the emotional wellbeing of the person performing the act, but also on others around that person. Thus every small act causes a small ripple and if there are enough small acts they could magnify into the butterfly effect of making people happier on the other side of the globe. The global project called Action for Happiness is based on the principle that kindness breeds happiness. It encourages people to perform small acts of generosity, from hugging to saying sorry or giving up a seat on a bus.

Myriam Mongrain of York University in Canada and her colleagues asked 700 participants to be nice to others over the course of a week. The participants were asked to act compassionately towards someone for five to ten minutes a day by actively helping or interacting with them in a supportive and considerate manner. When participants' levels of depression, happiness and self-esteem were reassessed after six months, they showed significantly a greater increase in self-esteem and happiness compared with those in the control group. The small acts of kindness not only made the participants happier, but the boost in mood stayed with them for months. 'The concept of compassion and kindness resonates with so many religious traditions, yet it has received little empirical evidence until recently,' remarks Mongrain. Does her empirical evidence prove the law of karma? You be the judge.

American researchers James Fowler and Nicholas Christakis followed the happiness levels of 4739 individuals from 1983 to 2003 'to evaluate whether happiness can spread from person to person and whether niches of happiness can form within social network.' Their results, reported in 2008 in the prestigious *British Medical Journal*, showed that 'people's happiness depends on the happiness of others with whom they are connected.' Their study provides proof that happiness spreads dynamically in a large social network.

Maja Storch, a psychologist at the University of Zurich, suggests two small acts that can create immediate happiness for you and those around you: (1) when you get to work greet co-workers with a 'good morning' before you check your emails; and (2) buy a flower during lunchtime and beautify your desk.

Let little wings of little acts flutter to create the butterfly effect.

36

That dreadful deadline

In his most famous novel, *Fathers and Sons*, Ivan Turgenev writes, 'Time, as we all know, is sometimes a bird on the wing, and sometimes a crawling worm.' He is describing time as we perceive and experience it in our lives. During a boring afternoon at work it seems to slow down like a crawling worm, whereas at an evening with friends it seems to fly like a bird.

We perceive time differently but how do we perceive it when a task has to be done? Gabriela M. Jiga-Boy of Swansea University in the UK studies the relationship between effort and time perception. In a series of studies she and her colleagues tried to find out the relationship between a task that had to be done by a certain deadline and how much time people perceived they had to do the task. Her results revealed how our mind translates effort into time:

When no deadline is imposed, events requiring more effort are perceived distant in time. It's our normal thinking: a difficult task requires more time to complete, so its completion must be further off.

When a deadline is imposed, events requiring more effort are perceived closer in time than events that require less effort.

Imposing a deadline reverses our minds' relationship between work and time, and we see a difficult task looming closer. A difficult task with an imposed deadline probably triggers emotions that skew our perception of the relationship between effort and time. Whatever the reason may be, 'much effort can signal little time, but little time may produce more adaptive action,' the study notes. Translation of the academic speak: if a deadline is imposed, we are likely to pay more attention to it and take action to complete it — within the deadline imposed.

Before you impose a deadline on finishing reading this book on time, spare a few more moments on the subject of time.

How do we perceive the past and future events? Logically, an unethical act performed yesterday should also be unethical if performed tomorrow. However, experiments by Eugene Caruso of the University of Chicago show that thinking about future events tends to stir up more emotions than events in the past. In general, we respond to future situations with increased emotions as a way to prepare ourselves for action. Even if we cannot affect future events, the emotional response persists.

As our emotions often guide our moral judgements, our judgements tend to be more extreme for offences that could happen in the future rather than those that occurred in the past. The same thinking also applies to good deeds: we tend to perceive good deeds more positively than bad ones. Taken together, the results 'demonstrate a systematic way in which moral judgements of the same action are inconsistent across time,' Caruso says.

Caruso's findings have practical implications in law and public policy. A pharmaceutical company, for example, may decide to go ahead with a new highly profitable drug even if research on its long-term safety is not conclusive. People's judgement (in the future) on this unethical action will be relatively less harsh because it will be about an event that happened in the past.

It is a complex and demanding task changing law and public policy. A deadline may help.

You may think you're not prejudiced, but you'll be horrified to find biases buried deep within your unconscious mind

37

Bias on the brain

The English novelist E. M. Forster once quipped, 'How do I know what I think until I hear what I say?' We do not know what is lurking in our minds until we speak our minds. Before 'we speak our minds', it's better we listen to psychologists and neuroscientists who in recent years have ferreted out an amazing array of stereotypical attitudes and beliefs about different races, genders, religions, sexual orientation, physical appearance and a host of other things that inhabit the minds of us all. Obviously, particulars vary depending upon our social and cultural environment; for example, Siri Carpenter, an American social psychologist, points out that in the USA about two-thirds of white people have an implicit preference for whites over black people, whereas black people show no average preference for one race over the other.

We all know when we encounter explicit prejudices and discrimination (redneck racism, ethnic stereotyping, and so on). But we don't know that we all harbour prejudices which we may even consciously loath. In one brain-imaging experiment to study such implicit or hidden prejudices, white participants were shown a series of pictures of white and black faces to measure activity in the amygdala, the region of the brain where fear is registered and generated. The experiment revealed a heightened activity when participants glimpsed black faces which were shown for only 30 milliseconds (too quick for participants to consciously process them). When faces were shown for half a second (enough time to consciously process them), the prefrontal cortex showed heightened activity. This region of the brain plays an important role in the ability to control thought and action with internal goals, showing that participants were trying to suppress their implicit anti-black bias. Other brain-imaging studies also show that people who claim not to be racist register skin colour automatically and unconsciously.

Project Implicit, a virtual laboratory for social and behavioural sciences at www.implicit.harvard.edu has an online test to measure your hidden biases. The test — it's free and everyone can try it — gives you an opportunity to assess your conscious and unconscious biases for more than ninety different topics, ranging from pets to political issues, ethnic groups to sports teams, and entertainers to styles of music. The test works by measuring the time a person takes to sort out 'good' and 'bad' words with pictures of different types of faces or objects. For example, in the weight (fat–thin) test, words and pictures of fat and thin people flash on screen in rapid succession. The test-taker links certain words and pictures with certain keys on the keyboard. The amount of time it takes a person to press the corresponding key is measured and then used to determine hidden biases. The weight test has been taken by millions of people around the world and the results are disturbing: they show an overwhelming automatic preference for thin people over fat people.

This innovative research project has not only revealed insights into our feelings and thoughts, but has also suggested that it may be possible to reduce racial bias. The idea that our neurons have plasticity is a new one. 'Any time you can get people to treat people as individuals, you reduce the effect of stereotypes ... It's remarkable that our brain is so flexible than ten hours of training will affect something that is the product of your whole life experience,' says Michael Tarr, an American cognitive scientist. This new research one day will lead to training for people working in potentially race-sensitive situations such as police officers, social workers and immigration officers.

Hidden biases can also colour our decision-making. American psychologist Eugene Caruso and his colleagues asked 101 Bulgarian university students to make a series of choices between prospective teammates in a trivia quiz. Each participant viewed profiles of two potential teammates that described each person's education, IQ and previous trivia game experience. The profile also included a photo of either a thin or an overweight person. The researchers discovered that although participants stated explicitly that weight had little or no impact on their decisions, they actually gave up about eleven IQ points to have a thin rather than an overweight teammate. 'There is a price to pay for biases that we may not be even aware of,' warns Caruso. The negative consequences of such biases are felt by more than just the target of discrimination. The decision makers in this experiment themselves paid a cost — 11 IQ points — for their bias, just as a hiring manager may pay a cost for hiring the less capable.

Awareness of stereotypes that others hold about the groups to which we belong can undermine learning and performance

38

Stereotypes sap success

A stereotype is an oversimplified picture formed from a single perspective. The picture, which is by its nature false, reveals ignorance often tainted by prejudice. We experience a 'stereotype threat' in a situation in which we fear that our performance will validate an existing negative stereotype about our social group, causing anxiety and therefore a decrease in performance.

Studies of students from ethnically and socioeconomically diverse areas show that students are aware of racial stereotypes early. In addition, stereotype threat leads students who are aware of broadly held negative stereotypes about their groups to perform poorly on standardised tests. Similarly, the common negative stereotype that girls aren't good at mathematics hurts girls' achievement in mathematics.

Stereotype threat blocks the path to learning by making students anxious. When we are tackling a problem, say an algebra equation, the brain processes the information through the amygdala, responsible for emotions. The amygdala then prioritises it to go through the prefrontal cortex, the 'decision-making' region responsible for memory and critical thinking. When we are under stress there is more activity in the amygdale than in the prefrontal cortex. Thus, even minor anxiety can block a student's ability to respond.

A US study demonstrates that the stress reaction may hit hardest the students who might otherwise be the most enthusiastic about mathematics. The researchers tested students on mathematics in either neural situations or in stereotype threat situations to invoke anxiety (by mentioning sex stereotypes about mathematics to women who were being tested). Another US study has found that women in similar stereotype threat conditions attempt fewer mathematics problems than men. However, various studies show that women can improve their performance in mathematics when they identify with women role models who are strong in mathematics or they read about successful women role models.

Robert J. Rydell of Indiana University, whose research focuses on stereotype threat involving women and mathematics, says that the effect on learning could be cumulative. 'If women do not learn relatively simple skills early on, this could spell trouble for them later on when they need to combine a number of simple skills in new, complicated ways to solve difficult problems,' he says. 'For example, if a young girl does not learn a relatively simple principle of algebra or how to divide fractions because she is experiencing threat, this may hurt her when she has to use those skills to complete problems in geometry, trigonometry or calculus tests.'

A study by Valerie Taylor of Princeton University and Gregory Walton of Stanford University suggests that negative stereotypes can prevent minority students from learning new academic material. But once the concerns about the stereotype have been alleviated, students' learning improves dramatically. The researchers also say that there are many ways to reduce the effects of stereotype threat on students, such as having role models who belong to a stereotyped group and promoting the mind-set that intelligence is malleable.

According to a team led by Alexander Haslam of Exeter University in the UK, although some stereotypes can promote failure, they can also lift a person's or group's performance and be tools that promote social progress. In an article in *Scientific American Mind* they cite research that shows that Asian women perform better on mathematics tests if they think of themselves as Asians rather than women. Another study cited by them indicates that if white golfers believe they are being compared with black golfers, they perform better if they think golf tests strategy but worse if they think it tests athletic prowess.

'We are not doomed to be victims of oppressive stereotypes,' Haslam and colleagues write, 'In short, who we are determines both how we perform and what we are able to become.'

Some people see sounds, smell colours and taste shapes — this mingling of two or more senses is called synaesthesia

39

When your senses give you more than you expected

'When I see equations, I see the letters in colors — I don't know why ... light-tan j's, slightly violet-bluish n's and dark brown x's flying around,' confides American physicist Richard Feynman in his book, *What Do You Care What Other People Think?*

In his memoir *Speak, Memory*, Russian novelist Vladimir Nabokov lists his own private alphabetic palette: the letter 'a' of the English alphabet evokes a sense of the tint of weathered wood, 'b' burnt sienna, 'c' light blue and 't' pistachio green.

In a letter to *Time* magazine, a mother described her daughter's full-colour alphabet: 'She sees numbers and letters of the alphabet in various colors. A is green, B is red, lowercase b is pink, 2 is yellow, 3 is green. 9 is orange and yellow. Months of the year also have their individual colors, which she notes each time she turns over a new calendar page.'

Like Feynman and Nabokov, the girl has a bizarre life-long condition known as synaesthesia in which sensory signals are mixed up. It doesn't only involve colours; in some people it involves sound, smell and other senses. And it's not a disability; it's a gift. Most synaesthetes say they enjoy the condition because it enriches their lives and some even feel sorry for those who do not have it.

Although known for about three centuries, synaesthesia (meaning 'joined sensation') was first described scientifically by Charles Darwin's cousin, Francis Galton, in 1880. There are no accurate estimates of how common it is. A study by psychologist Julia

Simner of Edinburgh University shows that synaesthesia appears in about one in twenty people (not one in 2000 as was the accepted wisdom until recently), and women are not more likely to have it than men. There is strong evidence that it runs in families (Nabokov's mother Elena was a synaesthete, as was his only child Dmitri; interestingly, so was his wife Véra).

If we calculate different combinations of the five senses — vision, hearing, touch, taste and smell — we would have twenty potential types of synaesthesia. But combinations are very large as a person may experience mix-ups of more than two senses.

One of the common types of synaesthesia is known as grapheme-colour synaesthesia in which people experience colours when they see particular letters or numbers and even think about them. Each person has their own specific associations and these associations are consistent throughout their life. A Canadian study of a girl doing arithmetic shows how automatic and involuntary these associations are. When the girl was shown a problem such as 5 + 2 =, followed by a patch of yellow (the colour she associated with 7), she named the colour and number faster than control subjects. But when she was shown a different colour, she took significantly longer to respond.

Another common form of synaesthesia is coloured hearing. The Russian composer and pianist Alexander Scriabin who died in 1915 was the first to rigorously catalogue his colour-note associations, declaring that C-sharp was violet and E-flat major red-purple. In other common types of synaesthesia spoken words produce taste, smells produce shapes or tastes produce colour sensations.

Gian Beeli and his colleagues at the University of Zurich studied a 27-year-old professional musician who has a unique condition of synaesthesia. She not only has the common tone-to-colour synaesthesia (for example, C-flat is red, F-sharp is violet), but also experiences a taste on her tongue whenever she hears a specific musical interval (for example, a sour taste for the tone interval minor second, sweet for major third, low-fat cream for major sixth, and no taste for an octave). Remarkably, she also makes use of her synaesthetic sensations in the complex task of tone–interval identification.

What causes synaesthesia? Neuroscientists dismiss the common explanations that it is fakery, or that people with this condition are merely being metaphorical (describing C-flat as red is like someone describing a tie as 'loud'), or that they are simply experiencing childhood memories and experiences. Brain-imaging studies show that synaesthesia is real; for example, the brains of coloured-hearing synaesthetes show activity in their colour centres in response to sounds. Perhaps it occurs when two separate areas of brains are linked. Research on synaesthesia is not yet conclusive. But it has already shown that not all synaesthetic experiences are caused by sensory inputs; some are even triggered by simply thinking.

You're indeed lucky if you possess this enhanced way of seeing the world.

The age-old heated debate over
birth order and intelligence continues

40
Blessings of birth order?

Are older siblings smarter than their younger brothers and sisters?

In 1874, Francis Galton, a man of many talents who coined the term 'eugenics' and the phrase 'nature versus nurture', was the first to answer this question. In his book, *English Men of Science: Their Nature and Nurture*, he advanced the idea that more firstborn sons were in prominent positions than could be attributed to chance. Since then, interest in the importance of birth order continues undiminished (Google Scholar search lists more than 81,000 scholarly articles on 'birth order').

A large-scale 2007 study — its sample size was 241,130 men aged 18 and 19 years — by Norwegian researchers Petter Kristensen and Tor Bjerkedal seems to settle the question: older siblings are indeed smarter. Their analysis of IQ scores and birth order of participants shows that firstborns enjoy a 2.3-point IQ advantage over the next elder siblings. The second sibling, in turn, is a point brighter than the third. (In case you don't know, IQ or intelligence quotient, is the ratio of actual mental age, as measured by intelligence tests, to the mental age that is normal for a particular chronological age — 70 per cent of people have an IQ between 85 and 115.)

As some children have different social and biological ranks in the family, the researchers distinguished between these ranks by examining IQs of second- and third-born brothers who had an older sibling (male or female) who died in infancy. The 2.3-point IQ difference, therefore, reflects how participants were raised, not their absolute birth order.

This IQ difference might seem modest; however, it can have far greater importance than most people realise because it translates into about a 30 per cent increased chance of a child getting into a good university, comments Frank J. Sulloway, an American psychologist who is considered a leading authority on birth order. Other research supports his assumption: firstborns are overrepresented in top US universities. He believes that IQ increases by one point for every rise in birth rank — the oldest of three will have an IQ about three points higher that the youngest sibling.

In 2010 Kristensen and Bjerkedal analysed a much larger sample of 392,969 Norwegians to compare birth order and intelligence within families and between families. This clever cross-sectional comparison of siblings from family to family– as opposed to the first study's comparison of brothers within the same family — shows the same birth-order pattern: firstborns score higher on IQ tests. The study also confirms the findings of the earlier study that social rather than birth order is associated with educational attainments. It also suggests that firstborn female children do moderately better on IQ scores than firstborn male children.

Although Kristensen and Bjerkedal provide some convincing data, no one thinks that the birth-order debate is finally over.

Critics of the relationship between birth order and intelligence argue that firstborns tend to hit the parental-attention gold mine. According to one estimate, American firstborns spend an average of 3000 more hours with their parents than their younger siblings, which accounts for higher scores on IQ tests. Parenting efforts does make a tremendous difference in a child's educational achievement.

Oldest siblings may have higher IQ scores than younger siblings; the difference in IQ scores is nothing to worry about if you're a younger sibling. You will find your own niche as psychologists assign other qualities to you: more tolerance for risk, and aptitude for becoming an artist, adventurer or entrepreneur.

41

The one and only

If you think only children are lonely, selfish, spoiled, arrogant brats, then think of Franklin Roosevelt, Winston Churchill, Cary Grant, Elizabeth Taylor, Elvis Presley, Frank Sinatra and John Updike. The negative labels attached to only children are nothing but unsubstantiated prejudices.

These stereotypes dates back to 1896 when American psychologist Granville Stanley Hall published a study, 'Of Peculiar and Exceptional Children', in which he described only children as lonely, spoiled, self-indulgent, bossy and generally maladjusted, and declared that 'being an only child is a disease in itself'. A decade later Austrian psychologist Alfred Adler echoed Hall's views when he said, 'The only child has difficulties with every independent activity and, sooner or later, they become useless in life.'

Since Hall no one has published research that shows what lies behind the negative stereotype of only children (even Hall's study was not scientific research in any sense; it was based on anecdotal evidence), yet we find it difficult to shake this stereotype of only children. It's a perfect example of how lies linger on in memory (*page 115*).

In recent decades, no one has studied only children — 'onlies' or 'singletons' — more closely than social psychologist Toni Falbo of the University of Texas at Austin. An only child herself and the mother of one, she has been researching only children since the early 1970s. She has found that only children are generally as well-adjusted, intelligent, accomplished and sociable as those with siblings. 'There is no evidence that supports the popular belief that only children are lonely, selfish or maladjusted,' she says. (She has spoken those three words so many times in the past 35 years that they run together as one: 'lonelyselfishmaladjusted', comments *Time* magazine.) 'Furthermore, some research evidence suggests that the absence of siblings can have beneficial effects on the development of children.'

Studies even show that the more siblings a child has, the worse they tend to do in school. Only children have a slight advantage in their academic achievement; for example, they may learn vocabulary early as they spend a lot of time talking to adults. 'The kind of person your child turns into has more to do with your parenting style and a whole variety of other factors than the number of siblings she has or doesn't have,' says social psychologist Susan Newman, author of *Parenting an Only Child: The Joys and Challenges of Raising Your One and Only*.

Research shows that there are no narcissist 'little emperors' in China, where more than 100 million children have grown up without siblings as the result of the government's one-child policy. Falbo, who has studied only children in China, says that whether in the east or west, only children do not show marked differences from children with siblings. China's only children are definitely not the arrogant centres of the universe that adults had feared.

Onlies are just like anyone else.

Recent work supports Darwin's idea that facial expressions are an innate code by which we understand each other's emotions

42

Faces are but a gallery of emotions

'Faces are but a gallery of pictures,' Francis Bacon wrote in his 1597 Essays. In 1868, Charles Darwin studied a gallery of pictures — eleven black-and-white photographs of faces, in fact — to support his research on the expression of emotion. The photographs were originally taken by a French neurologist G.B.A. Duchenne to examine the movement of facial muscles when an electric charge was directly applied to muscles. Darwin doubted Duchenne's conclusion — there are different muscles in the face that are responsible for every single, discrete emotion — and wondered whether there might instead be a smaller set of core emotions that are commonly expressed worldwide and across cultures.

Darwin is well-known for his theory of evolution but not for his theory that facial expressions of emotion are universal. He suggested in his 1872 book, *The Expression of the Emotions in Man and Animals*, that our ability to communicate subtle emotions with facial expressions may be innate: 'The inheritance of most of our expressive actions explains the fact that those born blind display them as well ... equally well with those gifted with eyesight.' For example, why is the raised upper lip included in one of the anger expressions? 'Darwin attributed this to it having been a "serviceable habit", exposing the canine teeth, which threaten harm to come, as well as preparing for the attack,' explains eminent American psychologist Paul Ekman.

Today, all evidence — from neuroscience, psychology and cross-cultural studies — supports Darwin's other monumental theory. 'To date, facial expression has been the richest source of information about emotions,' says Ekman.

Facial expressions are crucial for social communication and our brains are attuned to respond quickly to them. It takes a person only about 100 milliseconds to determine whether they like or don't like another person based on their face. However, if you take even longer to look at the face of Japanese person, you may not find out how the person is feeling. You are advised to pay attention to the person's tone, not the face as Japanese are taught to mask overt display of emotions. 'I think Japanese people tend to hide their negative emotions by smiling, but it's more difficult to hide negative emotions in the voice,' explains Akihiro Tanaka of Waseda Institute of Advanced Study in Japan.

As Botox paralyses facial muscles, it's difficult to read the emotions of people who use it. Now a study by University of Southern California psychologists suggests that Botox may also make the user less able to read the emotions of others. 'When you mimic, you get a window into their inner world,' says David Neal, the lead researcher. 'When we can't mimic, as with Botox, that window is a little darker.'

43

Living in a world of lies

Liars avoid eye contact.
Liars shift their postures.
Liars touch and scratch themselves.
Liars are nervous.
The speech of liars is flawed.

These are some worldwide stereotypes of liars documented by a group known as the Global Deception Research Team. In 2006 the team — it has ninety members, mostly psychologists, spread all around the world — carried out a study of stereotypes about liars in seventy-five different countries and forty-three different languages.

The team started with the hypothesis that there are cross-cultural stereotypes of liars, and the alternative hypothesis that these stereotypes are culture specific. They asked 2320 lifelong residents of fifty-eight countries the question: 'How can you tell when people are lying?' Responses to this question revealed several worldwide stereotypes, most importantly that liars avoid eye contact. This belief dwarfed other common stereotypes of liars, which included references to their nervousness, facial expressions, body movements and speech hesitations.

The team then asked 2520 lifelong residents of sixty-three countries ten simple questions to confirm these stereotypes:

When people are lying:

- do they act calm, nervous, or neither calm nor nervous?
- do they act silly, serious, or neither silly nor serious?
- are their stories more consistent than usual, less consistent, or neither?
- are their stories longer than usual, shorter, or neither?
- before answering questions, do they pause longer than usual, shorter, or neither?
- do they stutter more than usual, less, or neither?
- do they shift their posture more than usual, less, or neither?
- do they look at the other person's eyes more than usual, less, or neither?
- do they touch and scratch themselves more than usual, less, or neither?
- do they use hand gestures more than usual, less, or neither?

Responses to these questions confirmed that the strongest global belief is that liars avoid eye contact. This belief was present in every one of the seventy-five countries studied. Other stereotypes were also supported by the answers. Undoubtedly, there is a pattern of beliefs across the world, but these stereotypes are not necessarily accurate. Other research shows that the belief that liars avert their gaze has hardly any link with lying.

The Global Deception Research Team says that we have little ability to detect lies from behaviour when someone is lying. Many liars behave quite the opposite to what we expect. If we're bad at spotting liars, we can only hope that people who answered the team's questions were not lying.

44

Liar, liar, neurons on fire

Virtually everyone lies, we're natural-born liars. Yet we're not good liars — and hopeless at detecting liars. Contrary to popular belief, liars do not avoid eye contact, nor do they blink more or are fidgety (*previous story*). So where do these stereotypes come from? Aldert Vrij, a psychologist at the University of Portsmouth, points the finger at police manuals: 'They say that liars put their hands before mouths, look away, fidget and all these stereotypical types of behaviour, and there is absolutely no evidence for this — and neither do the manuals provide the evidence.'

If police manuals are of no help, is there a Pinocchio effect we can use to spot liars? One general difference Vrij has found between liars and truth-tellers is that liars blink less frequently and pause longer while speaking. Other psychologists suggest that emotions associated with lying — feelings of fear and guilt or delight at fooling others — can trigger a change in facial expressions. But these changes are so brief that most of us may never notice them without a good video recorder. Body movements and facial expressions may not provide clues to a liar, but what they are saying can. Liars tend to withhold information, either from guilt or to make it easier to get their stories straight, and repeat words and phrases, say American psychologists Bell DePaulo and Wendy Morris. 'Liars' answers sound more discrepant and ambivalent; the structure of their stories is less logical,' they say.

The stress caused by lying produces some involuntary physiological changes: more sweating, slow breathing and a brief drop in heart rate. These are also the symptoms of fear and anxiety. In 1921 John Larson, a University of California medical student, invented a machine (now known as the polygraph or lie detector) to measure these physiological changes. In a modern polygraph, rubber tubes placed around the chest and stomach measure respiratory rate, two small metal plates on the finger records sweat, and an electrocardiogram (ECG) registers the heart rate. During the test the operator asks a series of true–false questions related to a crime. Advocates of the polygraph test say that it is highly accurate. To critics it is as imperfect as measuring Pinocchio's nose. Why this sarcasm? The extremely artificial environment in which a polygraph test is performed makes subjects anxious and they experience sweaty palms and racing pulses.

Using fMRI scanners, neuroscientists have discovered that we use different parts of the brain when we lie and when we tell truth. Neurons in seven areas of the brain fire up when we tell a lie; neurons in only four areas fire up when we tell the truth. It seems that that the brain has to work harder to tell a lie. Mental conflict arises when we tell a lie and there is increased demand for motor control when suppressing a truth. No wonder neurons in your brain, not your pants, are on fire when you tell a lie.

Can 'innocent anxiety' confuse the fMRI? It's too early to say whether an fMRI scanner would make a perfect lie detector. Matthias Gamer, a German neuroscientist, believes that neither polygraphs nor fMRI can detect a liar with 100 per cent accuracy. 'Nevertheless, researchers may eventually identify a combination of brain images and signals from the body that comes much closer than do current methods to provide an accurate depiction of deception.' It's good news for police and lawyers.

And now bad news for those who have lost the ability to spot lies and sarcasm. We may be poor in detecting liars, but we all have some ability to pick up lies and sarcasm. If we're no longer able to distinguish among lies, sarcasm and fact, warns a new research study, we may be showing early signs of frontotemporal dementia, the most common form of dementia among people under 65. Patients with other types of dementia such as Alzheimer's disease could easily recognise a lie. 'We want people to recognise that these social lapses are actually a disease — parts of the brain are being eaten away,' advises Katharine Rankin, a neuropsychologist at the University of California.

It's difficult to erase rumours from
memory even after making subsequent corrections

45
Lies linger on in memory

Machiavelli, the 16th-century Italian exponent of the art of the politics of duplicity, certainly knew a thing or two about the impression of misinformation on people's memories when he said, 'Throw mud enough and some will stick.'

The sticking power of the Machiavellian mud of misinformation has now been enforced by psychologists at the University of Western Australia. They say that lies have a lasting impact on our memory and, despite the best efforts to correct wrong facts, they cannot be completely erased.

Psychologists asked 160 college students, who were randomly divided into groups of 20, to read aloud at their own pace a news report about a warehouse fire in which some of the information was false (fire was caused by volatile materials negligently stored in a closet). For some groups the information ended here, while other groups were given a later news report which corrected the false information by stressing that the closet was empty.

Some groups were also warned at the outset that they might be misled by the news story and should be careful about getting the facts rights. This was done to investigate what psychologists call the continued influence effect of misinformation: people believe in the original information even if it is shown to be wrong. Students in these groups were less likely than other students to make mistakes in remembering the facts, but they still erred sometimes. 'Despite best efforts to correct misinformation it can't be completely eliminated,' says Ullrich Ecker, the lead author of the study. 'Setting the record straight is almost impossible ... Our memory is constantly connecting new facts to old and tying different aspects of a situation together, so that we may still automatically draw on facts we know to be wrong to make decisions later.'

Do our strongly held beliefs and emotions have any impact on the continued influence effect of misinformation? While emotions have no significant impact, says Ecker, continued influence effect has been shown in many settings (the purported link between certain vaccines and autism, for example, or between Iraq and weapons of mass destruction). If you believe in something strongly and it's really important to you as a person you will cling to that no matter what, he stresses.

Neuroscientists Sam Wang and Sandra Aamodt, the co-authors of *Welcome to Your Brain: Why You Lose Your Car Keys But Never Forget How to Drive and Other Puzzles of Everyday Life*, say that because the source of information is forgotten, the message and its implications gain strength. This phenomenon, known as source amnesia, can also lead people to forget whether a statement is true. 'Even when a lie is presented with a disclaimer, people often later remember it as true,' they say. With time, lies become facts.

In brief, our brains hold on to lies even after they have been proved wrong; it occurs even if the retraction of lies is understood, believed and remembered. No matter what cock and bull story you tell, some lies will linger on in listeners' memories. It's important to get your story absolutely right in the first place.

46

Pulling the wool over our own eyes

The saying 'An error does not become a mistake until you refuse to correct it' became popular when in 1961 the US President John F. Kennedy used it in a speech after the failed Bay of Pigs invasion of Cuba. Kennedy also said, 'We're not going to have any search for a scapegoat ... the final responsibilities of any failure is mine, and mine alone.'

Most of us, unlike Kennedy, are reluctant to admit our failures and mistakes and look for scapegoat. We try to justify our stance even if it means telling lies and more lies. Self-justification is not merely lying to others, it's lying to ourselves. The eminent American social psychologist Elliot Aronson believes that self-justification is more dangerous and more insidious than explicitly lying because 'we are not even aware a mistake was made, let alone that we made it.'

The mind's mechanism behind self-justification is powered by cognitive dissociation: the conflict between two opposing cognitions arouses a contrary psychological stance — called dissonance — which in turn motivates activities designed to reduce dissonance. In other words, it's a state of mind in which we oscillate between conflicting ideas, attitudes, beliefs and opinions. According to Aronson, dissonance produces mental discomfort, ranging from minor pangs to deep anguish; people don't rest easy until they find a way to reduce it. This process of reducing dissonance revs up self-justification. For example, when a teacher exhorts her students to use reusable shopping bags because plastic bags harm the environment, students expect that the teacher always uses reusable bags. If a student finds the teacher outside a supermarket with plastic bags full of grocery, points at the bags and smirks, two opposite ideas would cause tension in the teacher's mind: 'I'm a respected teacher and believe in what I teach' and 'I carry my shopping in plastic bags.' Most probably the teacher would ease this momentary mental distress by telling a little lie, 'Oh, I left my reusable bags at home.' If the dissonance is between 'I'm a responsible person and know that drinking and driving is dangerous' and 'I drive after one too many drinks at the pub' you are forced not to drive after having been at the pub or to find flimsy excuses for driving after drinking. That's how cognitive dissociation drives self-justification.

Self-justification has its uses: it means we do not spend time worrying and fretting. 'But the reason self-justification is so dangerous is that in justifying the road we do take, we fall into the trap of justifying everything that flows from that first decision,' says Carol Travis, the co-author of *Mistakes Were Made (But Not by Me): Why We Justify Foolish Beliefs, Bad Decisions, and Hurtful Acts*, in which she and Elliot Aronson tackle the inner workings of self-justification.The mechanics of self-justification see to it that we become more and more enmeshed in our decision and less and less able to consider the possibility that it is wrong, she adds.

Confabulation — making up fictitious memories without a conscious effort to lie or deceive

47

'Honest lying'

In his famous book of clinical tales, *The Man Who Mistook His Wife for a Hat*, neurologist Oliver Sacks describes the story of Jimmie G., a charming, intelligent man who has lost the ability to form new memories. He 'cannot remember isolated items for more than a few seconds and has a dense amnesia going back to 1945' when he was a 19-year-old radio operator in the US navy. In 1975, Sacks recorded a conversation with Jimmie G., then a grey-haired 49-year-old man:

> 'What year is this, Mr G.?' I asked, concealing my perplexity under a casual manner.
> 'Forty-five, man. What do you mean?' He went on, 'We've won the war, FDR's dead, Truman's at the helm. There are great times ahead.'
> 'And you, Jimmie, how old would you be?'
> Oddly, uncertainly, he hesitated a moment, as if engaged in calculation.
> 'Why, I guess I'm nineteen, Doc. I'll be twenty next birthday.'

Here're two more real-life cases from the files of other doctors. When asked to move his left arm, a stroke patient suffering from paralysis of the left side of his body, replied with absolute conviction that he had no problem with his arm, despite the clear evidence he was unable to move it. Another patient, when asked about his surgical scar, explained that during the Second World War he surprised a teenage girl who shot him three times in his head, killing him, only for surgery to bring back him to life.

These patients are not knowingly telling lies. In fact, they are engaging in, what Morris Moscovitch, a Canadian neuropsychologist, calls 'honest lying': they are not aware that their statements are false. They don't know that they don't know what they are claiming.

We all love telling tall tales about ourselves, but some people do really believe their false stories. To some extent, we all confabulate when we try to rationalise our decisions or justify our options. But clinical confabulation is a type of memory problem.

Memories are not stored as whole representation of events but as the elements of events that are retrieved and reconstructed into a coherent narrative. Moscovitch explains it by an analogy: memories are stored like beads in a jar which are strung into a necklace at the time of retrieval; however, damage to certain areas of the brain impairs this retrieval. People who confabulate may have damage to the orbitofrontal cortex, which lies in the frontal lobe behind the eye sockets. This region of the brain plays a monitoring role in the production of thoughts into actions by sorting out what is real and relevant. The failure of this checking process in the brain makes people confabulate. Young children can easily be encouraged to tell tall stories partly because of their underdeveloped frontal lobes which fail to suppress irrelevant information.

In his book, *Brain Fiction: Self-Deception and the Riddle of Confabulation*, philosopher William Hirstein describes confabulation as a knowledge deficit. If the knowledge isn't there, something fills the gap. Disturbed by their uncertainty, confabulators fill that gap with fantastical tales.

48

Such an unreliable witness

If you happen to witness a violent street robbery, how likely is that you would remember it correctly when the police interview you? How old and tall was the robber, what colour hair and eyes, what was he wearing, what weapon was used?

The police and the courts rely so heavily on eyewitness testimony, it's crucial that the statements are absolutely correct. But memory is such an unreliable witness. Even if our memory is patchy, we do our best to weave a credible yarn. As mentioned in the previous story, our brains confabulate when we try to make sense of fragmentary information. A wealth of research has highlighted that eyewitness memories can be influenced by the questions asked by the interviewer. Questions can supply fragments of information that the eyewitness will include in the answer.

Research by UK psychologist Daniel Gurney shows that not only misleading verbal questions, but the interviewer's hand gestures can also make eyewitnesses believe they saw something they didn't. In one of the experiments, sixty-six participants individually were shown on laptop screens a short film of a knife attack, which was followed by an onscreen 'police' interviewer asking questions about the theft. Each of the participants was asked the same questions, but the interviewer varied his hand gestures as he spoke. For example, when asking, 'Did you notice any jewellery?' he performed a 'ring' gesture to the finger or 'watch' gesture on the wrist when he said the word 'jewellery'. The film showed the attacker wearing a ring but no watch. A significant number of participants reported information that was consistent with the 'ring' or 'watch' gesture they saw. By subtle manipulation of simple hand gestures, the experiment showed it was possible to make people believe they had seen something that wasn't there.

Once people have made the first statement, contrary to popular belief, retelling it doesn't improve their memory of the event. Psychologists say that the risk of inaccuracy increases the more the story is retold and discussed; witnesses who tell their stories many times become increasingly sure of their version of the details of the crime.

A study by UK psychologists has cast doubt on a technique used by police interviewers to stop eyewitnesses making details up. Coral Dando, the lead researcher, says the technique — asking eyewitnesses to say what they saw in reverse order — doesn't help in recall, instead it makes things worse. The researchers showed fifty-four participants a short film of a staged mobile phone robbery. The participants were then split into three groups, and two days later, interviewed about the robbery. In one group, participants were asked to describe the robbery in any order they remembered it, and then they were asked to describe it in reverse order. The second group was first asked to describe the robbery backwards and then freely. The third group, the control group, was asked to recall the robbery freely both times. The robbery film was divided into ten scripted actions in event order (the victim talking on the phone, the robber follows the victim, the robber grabs the phone, etc.). The results showed that the control group (with no reverse recall) averaged 48.9 correct event recalls about the robbery. The first group (free recall and then reverse recall) averaged 42.2, but the second group (reverse recall followed by free recall) did worst with a score of 38.7. This group also had the highest rate of confabulations (an average of 1.4 pure inventions per participant). The researchers concluded that reverse order recall has no positive benefits: witnesses can be made to forget as much as they can be helped to remember.

If eyewitness accounts are so murky, shouldn't they be trusted only when they have been independently corroborated?

Online infidelity is as damaging to relationship as 'real-life' infidelity

49

Cheating hearts in cyberspace

For whatever reasons researchers love to explore infidelity — emotional or physical intimacy outside a relationship. The wealth of their observations includes: people who engage in infidelity are more open to new experiences and are more extroverted than their partners; a low degree of agreeableness with partner, low conscientiousness and depression seem to be strongly related to infidelity; husbands are more suspicious of a wife's potential sexual infidelity and are more likely to discover a wife's affairs; men are more likely to engage in sexual infidelity, whereas women tend to have emotional attachments; male and female rates of infidelity are becoming increasingly similar particularly in younger people; and infidelity is more widespread than we think.

The internet has made infidelity much more widespread — it's only a few taps away on ubiquitous smartphones — and has opened a whole new world for researchers to explore. This new world has new 'online infidelity' words such as cyber affair, cybersex and most sexy of them sexting (sending sexually explicit messages or photos, mainly between mobile phones; the word first appeared in 2005 in the London-based Sunday Telegraph).

You may think that cyber affairs are harmless because no physical intimacy is taking place, but the consequences of becoming emotionally involved with someone 'online' — loss of trust or hurt — are as damaging to a relationship as 'real-life' infidelity. Women who value emotional attachments more than men see cyber affairs as cheating.

A study of online infidelity and sexting by American psychologists Diane Kholos

Wysocki and Cheryl Cilders shows that women are more likely than men to send sexually explicit text messages, often containing nude photos of themselves. 'Cheating is alive and well, and sexting is on the rise,' Kholos Wysocki says. 'Sexting is more than just sex. It's about feeling appreciated and good about yourself.'

Their research also reveals that men and women are equally as likely to have cheated both online and in real life while in a serious real-life relationship. In addition, older men are more likely to cheat in real life than younger men. Their research suggests the obvious: as technology changes, the way people find each other and the way they attract a potential partner also changes.

But the new technology of brain imaging supports the traditional view that the brains of men and women process cues of sexual and emotional infidelity differently. Men show greater activity than women in the brain regions involved in sexual or aggressive behaviours such as the amygdala and hypothalamus. Women, on the other hand, show greater activity in posterior superior temporal sulcus (this region lies in the top and back portions of the brain and is generally activated when the mind is trying to figure out social relationships).

Forget the technology. The ultimate truth is that humans are social creatures and are more interested in physical face-to-face contact than ethereal online contact. You haven't yet heard the last word on infidelity from researchers. A titillating topic worthy of investigation — and gossip (*page 154*).

Your self-esteem on
the cyberpsychologist's couch

50

Log onto Facebook to view your opinion of yourself

You might not feel better after staring at yourself in the mirror, but looking at your Facebook pages might enhance your self-esteem. This is probably because Facebook and other social networking sites allow you to choose what you want to reveal about yourself and filter anything that might reflect badly. Furthermore, feedback from friends posted on these sites tends to be overwhelmingly positive. Before you rush to log in, thank American psychologists Amy Gonzales and Jeffrey Hancock for their intriguing proposition: 'Unlike a mirror, which reminds us of who we really are and may have a negative effect on self-esteem if that image does not match with our ideal, Facebook can show a positive version of ourselves.' Hancock says. 'We are not saying it's a deceptive version of self, but a positive one.'

In their study, sixty-three university students, who were randomly divided into three groups of twenty-one each, were left alone in a media lab. One group was seated at computers that showed their own Facebook pages, the second group at computers that were turned off, and the third group at computers that were turned off but had mirrors propped against the screen. Those who were on Facebook were given three minutes to review their pages. Then all students were given a questionnaire to measure their self-esteem. The students who were on Facebook gave more positive feedback about themselves than the other two groups. Those who had made changes to their Facebook pages reported the highest self-esteem.

The results indicate that how we are able to present ourselves to others is important to self-esteem. By providing multiple opportunities for selective self-presentation — through photos, personal details and witty comments — social networking sites like Facebook can influence impressions of the self. 'For many people, there's an automatic assumption that the internet is bad,' says Hancock. 'This is one of the first studies to show that there's a psychological benefit of Facebook.'

To Canadian psychology researcher Soraya Mehdizadeh, Facebook is the online equivalent of gazing at your wall mirror. Those who spent more time updating their Facebook pages were more likely to be narcissists and lacked self-esteem, she says. After gaining access to the Facebook accounts of 100 college students she measured the frequency and duration of their activity on Facebook and actions such as photo sharing, wall postings and status updates. The participants took a psychology test to measure their level of narcissism.

Narcissists do not focus on interpersonal intimacy, but use relationships to appear popular and successful. They also inflate their sense of self-importance as a defence against feeling inadequate. Mehdizadeh thinks that Facebook allows narcissists to pursue an infinite number of virtual friendships and emotionally detached communication. Her study showed that narcissists were more likely to self-promote through status updates such as 'A girl should always be two things: classy and fabulous — Coco Chanel' or by posting photos entitled 'My Celebrity Look-alikes' which compares the photo of the user with that of celebrities. The study also showed that those with low self-esteem also checked their Facebook pages more frequently than normal.

Mehdizadeh supports the link between Facebook and self-esteem: social networking sites like Facebook are likely to have positive effects when used by people with low self-esteem or depression. 'I don't think it's necessarily a bad thing that people with low self-esteem use Facebook,' she says. However, she is not sure whether excessive time spent on Facebook by regular users can turn them into narcissists.

After studying typical behaviours of 311 college students on Facebook, American psychologist Michael Stefanone has found that women who base their self-esteem on their appearance tend to share five times more photos online than men. They also maintain larger networks on Facebook and other social networking sites. Mehdizadeh's study also reports similar findings: men generally promote themselves by written posts while women tend to carefully select the photos in their profile. Not surprisingly the results confirm the cultural focus on female image and appearance spills over from reality to virtual life.

Now it seems it also spills from virtual life to reality.

Brains of teenagers show
low motivation — it's just a phase

51

Confirming what parents already know

Everyone knows that teenagers are more willing to take risks than adults. An increase in risky behaviour is one of the most significant characteristics of teenagers. Brain-imaging studies of the adolescent brain now show why teenagers do crazy things.

The answer lies in a tiny region of the brain known as the ventral striatum. It's closely linked with the limbic system that controls emotions and motivation. It's also associated with anticipating rewards: it shows a surge of activity when something better than expected happens. But it shows less activity in teenagers than in adults.

While their brains were being scanned, twelve teenagers and twelve adults played specially designed gambling games, in which they anticipated the opportunity to obtain monetary gain or avoid monetary loss. The teenagers showed only about half as much activity in the ventral striatum as the adults. 'It appears that the brain circuitry in motivation to get rewards is under-engaged in teenagers and so it explains why they need extreme stimuli to achieve the same level of activity,' comments James Bjork of the US National Institute on Alcohol Abuse and Alcoholism who carried out this experiment with his colleagues.

The ventral striatum releases dopamine that helps the brain process rewards. Bjork suggests that teenagers seek more extreme behaviours to achieve normal levels of stimulation in ventral striatum. With more dopamine flowing, teenagers are likely to feel that risk behaviour is much more rewarding than it might seem to adults. No wonder teenagers take risks that don't seem worth it to adults — such as binge drinking, taking drugs, explicit sex-texting — because they require enhanced motivation.

It also explains why teenagers dislike household chores — not because they are slackers but because they are unmotivated. If you're a parent of teenagers, don't despair as it's only a passing phase. They might now be as mad as a hatter, but as they grow activity in their ventral striatum will also grow, making them as sound as bell.

52
What crisis?

In middle age, someone has said, life is restructured in terms of time left to live rather than time since birth. As birthdays become a stark reminder of time's passage, midlife (roughly from 40 to 65) also brings the expanding waistline and the feeling that our best years are behind us. The real and imagined implications of the emotional transition during this period of life have been called the 'midlife crisis'.

The idea of a midlife crisis began with Sigmund Freud who in 1907, when he was 51 years old, wrote, 'About the age of 50, the elasticity of the mental processes on which treatment depends is, as a rule, lacking. Old people are no longer educable.' Ironically, Freud himself produced some of his best works after the age of 65.

Research, however, doesn't support the views of Freud and his followers. David Almeida, a psychologist at Pennsylvania State University, believes that the midlife crisis is largely a myth. 'Midlife crises are often defined by someone else's perception rather than our own,' he says. There are a multitude of so-called midlife crisis experiences such as relationship problems or job dissatisfaction. 'Although it may seem easy to blame a midlife crisis,' he explains, 'age most likely has nothing to do with it.'

Laura Carstensen, a psychologist at Stanford University, agrees that people in the middle of their lives sometimes do have hard times, but they aren't more at risk of a crisis in midlife than other times in their lives. 'There is absolutely no empirical evidence for midlife crisis,' she says.

Neuroscience also rejects Freud's idea that 'old people are no longer educable'. It has been clearly established that far from remaining static in adulthood, as believed earlier, the brain continues to grow and develop throughout our lives. 'If you look at measures of knowledge like information tasks, vocabulary tasks, then these abilities seem to rise at least into the 50s and hold maybe into the 60s and 70s,' says Carstensen.

In contrast to grey matter which mostly contains neurons, white matter mostly contains closely packed axons coated with myelin. Axons — nerve fibres that carry a neuron's signal to other neurons — can be up to a metre or three feet long. Each axon is coated in a myelin sheath which acts like insulation around electrical wiring. Myelin helps transmit signals quickly. Myelin continues to grow well into middle age peaking at 50 on average, making the brain more nimble. There is also evidence that middle-age brain is also better at dampening emotions. As a result it's less and anxious and much happier brain. Not the grumpy brain of midlife crisis anecdotes.

Contrary to popular belief, large numbers of neurons do not die during adult years. There is no reliable physical evidence for greater number of dead neurons in otherwise healthy brains.

The brain's midlife crisis begins when myelin begins to breakdown. It's a slow process and usually revealed after 60. The loss of myelin disrupts the transfer of information through axons, which probably accounts for most of the age-related slowdown. Physical and mental exercise, good diet and social networking can help to maintain myelin and delay those 'senior moments', those irritating instances of forgetting such as where we left our keys last.

As far as 'midlife moments' are concerned, follow Almeida's advice: It's more useful for individual to focus on day-to-day challenges rather than ascribing them to an overall midlife crisis situation.

In the absence of an 'elixir of life', the best way to keep an old mind young is to exercise the grey cells regularly

53

Old age: betrayal by our own cells

Jeanne Calment, the Frenchwoman who died in 1997 at the age of 122, was probably the oldest living person ever. The chances of you and me living to her age are similar to our chances of swimming like Michael Phelps and win eight gold medals in a single Olympics.

If we can function well in youth, why can we not continue to do so in old age? Why do humans — and, for that matter, all organisms — age?

The answer to these questions lies in the genes. If we want to beat Madame Calment's record, choosing the right ancestors would be a big help. They should be rich (wealth brings health) and should have right genes (not the genes with predisposition to disease such as cancer or Alzheimer's).

Even if we have the right genes, they cannot help us to live forever. Cells are damaged all the time, sometimes they began to replicate uncontrollably and become cancerous. This damage is accumulated over time leading to the breakdown of healthy functions of the body. To help us live longer and healthier scientist are trying to find ways to manipulate damaged cells. An anti-ageing pill is not yet on science's horizon, but it hasn't stopped bogus health gurus and snake oil salesmen to peddle anti-ageing 'therapies'. Scientists warn that no anti-ageing remedy on the market has been proven effective. However, a large body of research suggests that a low-calorie diet — restricting calories below normal consumption levels — can slow down ageing.

As far as the brain is concerned, there are ways to keep it active in old age without any help from anti-ageing hucksters. The brain starts losing cells or neurons from the age of thirty, but don't be alarmed as you have 100 billion neurons. And when a neuron dies, neighbouring neurons starting sprouting new connections and, within limits, restore the brain to its older self. In some instances the brain does make new neurons and rewires itself, and some scientists now dream of drugs that would stimulate the brain to replace its own neurons. Not everyone agrees with their dream; the best strategy is lifestyle changes: a good diet, lots of sleep and physical exercise can increase the number of neural connections better than any drug.

Short-term memory declines with age. You'll know when you begin to write Post-it notes to yourself or begin to mutter whatshername is on the tip of your tongue but can't recall. At around age 70, memory and reasoning ability begin to decline. Your doctor will tell you that this decline is the ageing brain's equivalent of creaky knees. There are no pills yet to stop this decline, but you can slow it down by exercising the grey cells, the closely packed neurons that form the brain's grey matter. Exercising is simple: read, learn new things, don't be a loner or a couch potato — interact with people, and smile and laugh (Madame Calment credited her long life to her sense of humour, 'I'll die laughing,' she said).

Why wait to exercise the little grey cells? You can start right now. How many words in this book? Answer it the way Enrico Fermi, the greatest Italian scientist of the twentieth century, would have answered: by making reasonable assumptions, not necessarily relying upon definite knowledge for an exact answer. Fermi revelled in posing unexpected questions on aspects of the natural world and then figuring out their answers (for example, how many atoms could be reasonably claimed to belong to the jurisdiction of the USA?). Now such questions are known as Fermi questions.

Here're million ways to order lazy neurons in your brain to jump and start making fast connections with their neighbours: Can you live to be a million hour old? How long would it take you to count to a million? How high are million you standing on each other's shoulder?

54

Loneliness is bad for you

So lonely am I
My body is a floating weed
Severed at the roots.
Were there water to entice me,
I would follow, I think.
Ono no Komachi*

Feelings of loneliness are quite common among older people. When loneliness becomes a chronic condition its effect on health can be more serious. These harmful effects include higher levels of anxiety, negative mood, dejection and stress; and physical effects such as increased risk of high blood pressure, stiffening of arteries, infections and impaired sleep. Some studies have even linked loneliness to changes in the brain and triggering a change in gene activity.

Social psychologist John Cacioppo of the University of Chicago who is renowned for his research on loneliness says that its effect can be seen on the brain. He is the first to use brain imaging to study link between loneliness and activity in the brain. He has found more activity in a region of brain associated with social stimulus and feelings of love in non-lonely people than in lonely people when they viewed pictures

*A Japanese poet who flourished around 850 AD. Legend has it that she died old, destitute and lonely. Her poems appear in 'Kokinsh', the first anthology of Japanese poems compiled in early 900s (poem translated by Donald Keene

of people in pleasant settings. In contrast, another region of the brain associated with understanding the perspective of another person is less active in lonely than in non-lonely people when viewing pictures of people in unpleasant settings. 'Given their feelings of social isolation, lonely individuals may be left to find relative comfort in non-social events,' he says.

Cacioppo's research has also shown that lonely people have more of the hormone adrenaline (also called epinephrine) flowing in their bodies. It is a 'fight or flight' hormone and high levels indicate that the lonely people have to go through life in a heightened state of arousal. As stress hormones are involved in fighting inflammation and infection, loneliness contributes to wear and tear of the body. Good sleep has calming influence on stress, but Cacioppo's research shows that lonely nights are disturbed by 'micro awakening' resulting in poor-quality sleep. Since sleep tends to deteriorate with age, lonely people's impaired sleep slows down body's natural restoration process. Poor-quality sleep has been linked to weight gain and diabetes.

Steve Cole of the University of California and his colleagues (including Cacioppo) have found that chronic loneliness triggers a change in gene activity. People who experience chronically high levels of loneliness show gene-expression patterns that differ markedly from those who do not feel lonely. This result suggests that loneliness is linked to changes in the body's immune system.

It's not all bad news. Loneliness can be overcome. Cacioppo says that the power of face-to-face interaction is fundamental to what we are; its richness affects our brains. 'The degree of social connection that can improve our health and out happiness ... is both as simple and as difficult as being open and available to others,' he writes in his book, *Loneliness: Human Nature and the Need for Social Connection*.

A study by psychologists from the Universities of Southampton in the UK and Sun Yat-Sen University in China suggests that for lonely people, drawing on nostalgic memories of happier times could provide a coping mechanism for their feelings, magnifying perceptions of social support and restoring an individual's feeling for social connectedness. The researchers believe that nostalgia, a sentimental longing for the past, is a psychological resource that protects and fosters mental health.

Another 'psychological resource' is junk food. A University of Buffalo research team has found that comfort foods really can be good for your heart and emotions. Comfort food can serve as a readymade, easy resource for remedying a sense of loneliness, the team recommends. The idea for this 'bonding with comfort food' study came from the researchers' observation that some people counteract loneliness by bonding with their favourite TV show, building virtual relationships with a celebrity or a movie character or looking at pictures and memories of loved ones. Too much comfort food may clog your arteries.

An activity which won't clog arteries and will provide ample opportunities for vital face-to-face interactions is volunteering for a cause in which you believe (*next story*). Volunteering gives older people a perfect opportunity to leave their 'homes full of silence':

> Men and women who are not yet done with being ferocious and bright but for whom time now stands empty as they wait in homes full of silence; their only misunderstanding to have lived to an age when they are no longer coveted by a society addicted to youth.
>
> – Ishani Kar-Purkayastha, 'An epidemic of loneliness', Wakley Prize Essay, *The Lancet*, 18–31 Dec 2010

Protective effects of
volunteering on health and wellbeing

55

Help others to live longer

Seniors who spend about two hours a week volunteering help themselves as well as others. Studies of older adults have established that people live longer because they volunteer, rather than that people volunteer because they're healthier and hence more likely to live longer. Undoubtedly, better health leads to continued volunteering which leads to improved physical and mental health.

What factors create the protective effects of volunteering? One possibility is that volunteering provides meaning and purpose in people's lives — and a sense of belonging (which is important for older adults as they are prone to social isolation). Such qualities may in turn lead to better health, greater happiness, less stress and depression and a little better sleep.

'One of the best ways to get your mind off your aches and pains is to get your mind on somebody else,' advises American ethicist Stephen G. Post, the bestselling author of *The Hidden Gifts of Helping*. 'Helping is a buffer against helplessness, and an affirmation of self-efficacy — I can do this!' The relationship between volunteering and health is rooted in biology. This is a new science, Post says, but it does seem that it involves the brain, it involves the immune system, and it probably involves certain hormones, like oxytocin — the compassion hormone. Volunteering also elevates the levels of the body's endorphins and dopamine, the neurochemicals that affect your mood.

To experience health benefits of volunteering, you have to be within 'volunteering threshold', meaning you have to give, on average, 100 hours a year to charity. If you volunteer for more time than this, you would not necessarily gain greater health advantages. Volunteering may benefit people of all ages, but adults 65 and older are most likely to receive health benefits.

Volunteers benefit most when the programs they are working for provide good 'organisational support'. This means greater choice of volunteer activities, training and ongoing support, and volunteers finding their work interesting, feeling that they are being used effectively and receiving positive feedback. All volunteers have rights and obligations. They are free to determine the time they would devote to voluntary work. They are also entitled to receive appreciation and recognition of their work. At the same time they must aspire to excellence in the task they take upon themselves. They must show the same professionalism they show in their (past or present) paid employment.

The essence of volunteering distilled by a slew of research is straight: by helping others you help yourself; but the mathematics is skewed: give a little, get a lot back.

The Motley Mind

Everyone yawns,
but no one knows why

56

Bored already? You may yawn now

Read or think about yawning and in a few seconds, it will make you yawn — for six seconds only. That's the life span of the average yawn. In everyday life yawning is a sign of boredom, fatigue or drowsiness. It's more common during the hour before bedtime and the hour after waking. We do not yawn during sleep, but rare cases of people yawning during sleep have been documented. We usually yawn when we stretch, but we don't always stretch when we yawn. You cannot yawn on command; it's spontaneous and unstoppable. It's also 'contagious': about 55 per cent of people will yawn within five minutes of seeing someone else yawn (we're not alone, chimpanzees, monkeys and dogs also emit contagious yawns). These are some of the facts about yawning.

Are you yawning yet? No, then read on. Why do we yawn? American psychologist Robert Provine is a leading authority on this common, if not always appropriate, involuntary behaviour. He has spent years studying yawning but hasn't yet found a Grand Unifying Theory of Yawning. No, the reason is not that he reads and thinks so much about yawning that he is probably yawning all the time and can't concentrate on his research. 'I see much potential in using yawning to develop and test theories of mind,' he says.

He is sceptical about the commonly held beliefs about yawning that it occurs as a response to a lack of oxygen or too much carbon dioxide in the blood, or a need to increase circulation in the brain. His experiments have confirmed that even breathing

100 per cent oxygen did not inhibit yawning. At the same time the experiments have also rejected the ancient theory of Hippocrates that yawning got rid of 'bad air' and increased 'good air' in the brain.

Other scientists have shown that paraventricular nucleus of the hypothalamus seems to be the 'yawning centre' of the brain. It contains chemical messengers such as dopamine and oxytocin which can induce yawning. Yawning may also be a means of communicating to other changes in environmental or internal body conditions, perhaps to synchronise behaviour. If this is the case, American physiologist Mark A. W. Andrews says, yawning in humans is most likely an evolutionary ancient mechanism that has lost its significance.

Provine believes that yawning is a social cue to switch activities: 'I'm bored. Let's talk about something else.' Point taken. Let's move on to the next topic, 'Chucklers love nutty chocolates', a sweet and jolly topic that's sure to kill your yawn.

Laugh all the way
to the chocolate shop

57

Chucklers love nutty chocolates

The way you laugh can tell your friends about your personality. Researchers have matched seven broad categories of laughing style — snorters, belly busters, chucklers, sniggerers, howlers, gigglers and cacklers — with distinct character traits. While you are likely to be in good company with kind and thoughtful chucklers, flirtatious gigglers and sociable and trustworthy belly busters, you may well give snooty snorters, churlish cacklers, immature and insensitive sniggerers and attention-seeking neurotic howlers a wide berth.

Do you love nutty chocolates? People who choose chocolates with nutty centres are laid-back, but methodical, tend to think first then act (the kind of people who're reading this book). This finding is the result of a poll of thousand chocolate lovers (if you want to take the poll seriously, eat a chocolate bar before reading on).

Partygoers go for very sweet centres, shy people bitter-sweet centres and perfectionists and thinkers coconut centres. Coffee-flavoured chocolates lovers let others make decisions, while fudge lovers are easy going, sociable and act first, think later.

Shapes are important too: sociable people prefer circles and diamonds, relaxed types opts for ovals and the shy ones go for rectangular chocolates.

Does the chocolate research tally with the laughter research? A sweet and jolly question for you to answer, but don't let it drive you nuts.

While we're talking about chocolates, brains of chocolate lovers lit up more strongly when they see or taste chocolates than those who do not crave chocolates. The regions of brain thought to be important in habit-forming behaviour and addiction showed more activity in chocolate lovers when psychologists at the University of Oxford scanned the brains of chocoholics and non-chocoholics after presenting them with appetising pictures of chocolate bars. After seeing the pictures, the volunteers (all women) were allowed to taste liquid chocolate fed to them through a tube in the confined space of the fMRI scanner.

You may or may not be addictive to chocolates, but there is no link between chocolate consumption and happiness. At least, according to a study — 'research of a holiday kind' by paediatric physician Kevin Chan — published in the journal of Canadian Medical Association. Chan might have failed to notice rising happiness in his 180 study participants as they consumed chocolates; however, he was left with a year's supply of chocolates in his basement. That's happiness.

Is there a link between other Cs and happiness: candy, chips, cookies, caffeine and cola? The search for the answer will certainly make you chubby, if not cheerful.

The length of the index finger divided by
the length of the ring finger says a lot about you

58

If it's not in your mind, it's in your hands

Take a ruler, stretch your right hand palm up and measure the length of index and ring fingers, starting from the crease nearest your palm to the tip of the finger. Or, you can take measurements from a photocopy of the hand. Now divide the length of your index finger (second digit, 2D) by the length of your ring finger (fourth digit, 4D). The result is known as second-to-fourth-digit-length ratio (2D:4D), or simply the finger ratio. In most people both hands have slightly different finger ratios. The average ratio for women is above 1 and for men less than 1.

In the early stages of pregnancy, the womb is washed over by sex hormones oestrogen and testosterone. If you were exposed to more oestrogen than testosterone in the womb, the index finger will be longer than the ring finger. If you were exposed to more testosterone than the ring finger will be longer.

According to American anthropologist Helen Fisher, your finger ratio says something about your personality. If you have a longer index finger, 'you have good verbal skills, can find the right word rapidly, are good at remembering, better at compassion, nurturing, patience, have good people skills.' If you've a longer ring finger, 'you tend to have poorer social skills but be direct, decisive, ambitions, competitive'. Both men and women with a low finger ratio tend to be assertive.

Our early exposure to oestrogen and testosterone has an effect on how our body develops. There is growing evidence that it may affect predisposition in later life to disease and sexual orientation. Scientists have discovered a link between prostate cancer and finger ratio: men with higher ratios run a significantly higher risk of prostate cancer. Women with lower ratios are more susceptible to osteoarthritis and cervical cancer. Gay men seem to have significantly higher ratios than heterosexual men, and lesbian have a low ratio. This indicator is not strong enough to allow you to jump to conclusion about someone's sexual orientation by just looking at their fingers.

Korean researchers have stretched the 'power of prediction' of finger ratio a bit too far by linking it to penis length. They have found that men with a lower ratio tended to have a longer penile length. Their result is based on the study of 144 men twenty or older who were hospitalised for urological surgery. Their results: the average flaccid and stretched penile lengths were 7.7 cm (3 inch) and 11.7 cm (4.6 inch) respectively, while the average finger ratio was 0.97. Don't rush to generalise these results. The Korean study was conducted on a single ethnic group of men, and digit ratio has been shown to vary among ethnic groups.

And for trivia fans, Casanova believed that the ring finger is relatively longer than the index finger for both men and women. Those with a longer ring finger have a masculine finger ratio of less than 1. This is called the 'Casanova pattern'.

You may still believe in 'blue Mondays', but researchers have consigned this belief to folklore

59

Repaint your Mondays in a lighter shade of blue

'Monday's a blue day,'
someone said;
'Somber as storm clouds,
heavy as lead,'
But I have known Mondays
bright as a bell.

You may hate Mondays but researchers now agree with American poet R.H. Grenville's sentiments in his 1957 poem 'Blue Monday'. Your Monday moods may not be as 'bright as a bell'; they are no worse than those on Weekdays. Large-scale studies of mood, stress and physical illness on middle-class white-collar American workers show that positive moods are higher on Fridays, Saturdays and Sundays. Mood worsens on Mondays and stays about the same up to, and including, Thursdays. After all, American blues guitarist 'T-Bone' Walker may have a point when he crooned in 1947:

They call it stormy Monday
But Tuesday's just as bad
Wednesday's worse
And Thursday's also sad.

So, why is Monday viewed as grim and unpleasant? May be it's just the contrast with Sunday — the idyllic day — that produces perception that Monday is the worst day of the week. Or, may be blue Monday is primarily a result of the expectations imposed by the cultural myth. This cultural myth seems to have a long history ('Monday morning found Tom Sawyer miserable,' Mark Twain wrote in 1876. 'Monday morning always found him so.').

An internet survey of weekly mood cycles by Charles S. Areni of the University of Sydney has revealed that day-of-the-week cultural myths ('Monday blues', 'Wednesday hump day', 'Thank God It's Friday') were pronounced when participants *predicted* their moods for each day of the upcoming week, less obvious when they *remembered* their moods from each of the preceding week, and least apparent in the *momentary* moods they actually experiences on each day. These results suggest that the predicted and remembered moods were driven by the day-of-the-week stereotypes.

How could you turn your stereotypical Monday into as 'bright as a bell' day? The best way to prevent post-weekend depression is not to worry about Monday on Sunday. Before you get out of bed on Monday morning, spend a few moments thinking about what you want to accomplish on that day. Then you're most likely to be in tune with Grenville that blue is 'the loveliest color a day could be'.

Solace for spurned lovers, and some mathematics for yet-to-be spurned lovers

60

Why it hurts like hell when you're dumped

When American anthropologist Helen Fisher thinks about love and romance she pictures dopamine and serotonin, not flowers and chocolates. She likes to probe the human brain with the fMRI scanner to discover the chemical and biological nature of romantic love.

She studies the levels of dopamine and serotonin in the brains of 'men and women who had just fallen madly in love'. Dopamine and serotonin are neurotransmitters, chemicals that neurons in the brain release to send messages to each other. Dopamine produces exhilaration, increase energy, even mania — the core feelings of romantic love. Serotonin regulates mood, appetite, sexual desire, aggression and sleep. When her love-struck subjects look at a photograph of their beloved projected on a screen just outside the fMRI scanner, their brains show increased activity in areas that make and distribute dopamine. This suggests that intense, early stage romantic passion is associated with brain areas rich in dopamine. She concludes that 'romantic love is not, in fact, an emotion, but primarily a motivational state designed to make us pursue a preferred partner'.

'Indeed, romantic love appears to be a drive as powerful as hunger,' she says. 'No wonder people around the world live — and die — for love.'

What happens if the 'preferred partner' jilts you? After studying people who were suffering the trauma of a recent rejection in love, she says that we can only blame the wiring in our brain and the harsh realities of evolution, which ensure that rejected lovers can extricate themselves and start again. Falling in love profoundly affects our social and genetic future, she says. 'As a result, we are built to suffer terribly when love fails — first to protect the departure and try to win the beloved back and later to give up utterly, dust ourselves off and redirect our energy to fall in love again.'

Spurned lovers are advised to delve deeper into Fisher's book, *Why We Love: The Nature and Chemistry of Romantic Love*. If you've not been dumped (yet), science — well, it's not brain science — can help you to find out the future duration of your current relationship.

American astrophysicist J. Richard Gott is an expert on black holes. He also loves using 'Copernicus principle' to forecast the future duration of events as mundane as a couple's current relationship or as momentous as the longevity of the human race. In 1543 Polish astronomer Nicolaus Copernicus rejected the prevalent belief that Earth stood still at the centre of the universe. The 'Copernicus principle' says that one's location is unlikely to be special, or to put it bluntly, wherever or whenever we are, it's nothing special.

The mathematics of Copernicus principle, according to Gott, works as follows: 'If there is nothing special about your observation of something, then there is a 95 per cent chance that you are seeing it during the middle 95 per cent of its observable lifetime, rather than during the first or last 2.5 per cent.' Or, the future of a thing is between 1/39 and 39 times as long as its past.

In 1969 Gott visited the Berlin Wall which was then eight years old. He applied the above formula (with 50 per cent likelihood; in a later version he changed this value to a greater certainty of 95 per cent) and predicted it would last more than two and two-thirds years but less than 24. The Wall came down 20 years later in 1989.

Say 'wow' and apply Gott's mathematics to work out the duration of your current relationship. No, you cannot use it to calculate the duration of a relationship's future if you are reading this on you first date. You're at the beginning of an event.

The brains of adult males and females are not exact copies of one another, but the tiny differences do not dictate cognitive abilities

61

The myth of 'his' and 'her' brains

Since psychotherapist John Gray proposed his seductive idea of 'planetary differences' between the sexes — *Men Are from Mars, Women Are From Venus* — in his famous book in 1992, it has become popular to think of men and women fundamentally different in terms of their brains and behaviour. Is 'his' or 'her' brain the right indicator of behaviour when there is so much inconsistency within each sex? One needs to be sceptical about differences between men's and women's brains reported in the popular media, as they reinforce old-fashioned gender stereotypes.

The results of brain-imaging studies may appear real and reliable, but these studies are usually done on very small samples because the machines are complex and expensive. Scientists are aware of the limitations of their studies, but in the hands of popular media the results become 'scientific proofs'. In her book, *Brainstorm: The Flaws in the Science of Sex Differences*, Rebecca Jordan-Young, a sociomedical scientist, argues that most studies on innate sex differences in the brain are riddled with 'weaknesses, inconsistencies and ambiguity'. Neuroscientist Lise Eliot echoes the same disenchantment with such studies in her book, *Pink Brain, Blue Brain: How Small Differences Grow Into Troublesome Gaps — And What We Can Do About It*. She says that stories of 'sex differences in the brain are sexy', but assertions of sex differences are either 'blatantly false', 'cherry-picked from single studies' or 'extrapolated from rodent research'.

You cannot tell male and female brains apart just by looking at them. There are subtle differences. Men's brains are slightly bigger than women's brains overall, even when the size is corrected for the fact men are on average taller than women. As brain size is no indicator of intelligence, men and women show no consistent difference on IQ tests. Some anatomical and physiological differences also occur in male and female brains. For example, brains of males produces neurotransmitter serotonin at a faster rate than those of females. Serotonin influences mood, so it may explain why women are diagnosed with depression twice often as men.

However, there is no convincing evidence that these differences result in differences in cognitive abilities. Whatever the differences are, they are biological; they are not innate or hardwired. Now we know about neuroplasticity and how experience can change the brain. 'Because gender traits are influenced more by a partiality for typically masculine or feminine interests, jobs and fashion styles than they are by biological sex, it follows that our malleable brains reflect these experiences,' says Eliot.

Though there are some behavioural differences between the sexes (*next story*), they have been overblown, mostly by some educators. One of these differences is spatial reasoning: men show some advantage at mental rotation, which makes it easier to compare two-dimensional or three-dimensional shapes when rotated in space. Put simply, men are better at reading maps. But many wives will testify: they are better skilled than their husbands in reading maps. Some studies have reported that women are more likely to use landmarks and men more likely to uses distances and geometry when navigating. Differences in spatial ability have not been noted in young children, and studies show that spatial ability can be improved with training. There is no 'scientific proof' that says girls cannot successfully pursue careers in mathematics, physics and engineering.

Women also show an advantage in language on average (many studies and endless text-messaging by girls seem to prove it). But again, this difference hasn't stopped men becoming great writers. 'Kids rise and fall according to what we believe about them,' notes Eliot.

It's time to ditch old stereotypes of who's good at what.

62

Women are better at remembering faces

Episodic memory is a type of long-term memory based on personal experiences. More accurately, it is conscious recollection of personal experiences in terms of their content (what), location (where) and time (when).

Women are better than men in verbal episodic memory tasks like remembering words, objects, pictures or everyday events. Not surprisingly, they excel when asked to recall events that happened during the past year, day or minute. Men outperform women in episodic memory tasks that involve visual and spatial awareness (in real-life situation this male advantage means remembering way out of the woods, but not necessarily from the museum to the hotel in a strange city as more verbal information is available in the city). However, women are better at tasks requiring remembering an object's position such as remembering where they last saw the car keys.

Now, Swedish psychologists Agneta Herlitz and Jenny Rehnman have discovered that women are better on average than men at remembering faces, especially female faces. They have found this advantage across age: both young girls and adult women remember more female faces, irrespective of the age of the faces. 'The reason seems to be that females allocate more attention to females than males,' they say.

Canadian psychologist Jennifer Steeves has added a twist to the Swedish finding. She has found that gay men can recall faces faster and more accurately than straight men. When memorising faces gay men — and straight women — use both sides of the brain, while straight men tend to favour right side of the brain for such tasks. She has also found that left-handed heterosexuals can remember faces better than left-handed homosexuals, and also do better than right-handed heterosexuals. Left- or right-handedness is thought to be linked with sexual orientation; other studies have shown that homosexuals are 39 per cent more likely to be left-handed.

Talking of faces, a study from neuroscientists at Massachusetts Institute of Technology shows that we tend to remember pictures of people much better than wide open space. Most memorable photos are those that contain people, followed by static indoor scenes and human-scale objects. Beautiful landscapes are easily forgettable. 'Say cheese' when you are taking your holiday photos.

63

Pssst ... gossip changes the way you see a person

Gossiping may be an evil pleasure (a biblical sin against charity), but pleasure it is. It's good for your brain as well. Research shows that chattering or gossiping with your friends for just ten minutes improves memory and mental performance.

Gossip also gives us clues on whom to make friends with without spending lots of time with them first. It literally affects the way we see each other; it forces the brain to focus on people who might be threatening. These findings come from a study led by cognitive neuroscientist Lisa Feldman Barrett of Northeastern University in Boston. 'Encountering negative gossip about someone makes it easier to register that person's face than neutral or positive gossip does,' she says.

Participants in the study were first shown neutral faces paired with negative gossip (described in written sentences such as 'threw a chair at his classmates') positive gossip ('helped an elderly woman with groceries') or neutral gossip. The brain usually combines two images into a single image. However, if one image is shown to one eye and a different image is shown simultaneously to the other eye, the visual information does not match. Instead of seeing two superimposed images, we only see one image at a time — one image is seen for few moments, then it changes into the second, then it reverts to the first, and so on. This phenomenon is called 'binocular rivalry'. After seeing the face-sentence pairs four times, participants now looked into a binocular-rivalry device that presented the images of the faces (they had previously learned to associate with negative, positive or neutral gossip) to one eye and one unrelated image (in this case, a house) to the other eye. By pressing computer keys, participants were able to record the duration they consciously experienced seeing a particular face or the house. In both experiments, participants saw faces linked to negative gossip for a longer time than any others.

Mild electric shocks can cause people to associate negative emotions with a face, but it can take dozens of repetitions. In contrast, Feldman Barrett says, people learn to associate negative gossip with faces very quickly, which shows you the power of gossip. We all know that what you see influences what you feel. Barrett's findings show that what you feel about someone influences what you see visually.

The findings also confirm that our 'sinful' interest in gossip is an evolutionary adaptation as it serves a protective purpose by helping us to predict who is friend or foe without first-hand experience of that person. They are also consistent with other evidence that mood can affect firing of neurons in the visual cortex, the region of the cerebral cortex that processes visual information. Hearing gossip about someone can change the way you see that person — sounds absolutely right.

64

Beauty is in the brain of the beholder

Every face doesn't launch a thousand ships; but if a face is attractive, your brain will know it even before you're aware you have seen it. Our brains' beauty bias was discovered by US psychologists Ingrid Olson and Christy Marshuetz in an experiment in which participants were presented with attractive and unattractive faces for only 13 milliseconds. They were much quicker to judge attractive faces than unattractive faces, showing that facial attractiveness can be assessed from brief glimpses of visual information. This finding makes it difficult to ignore beauty.

The saying that beauty is in the eye of beholder has been given a new meaning by findings of a study by a team of eleven scientists lead by Camilo J. Cela-Conde of Universidad de las Islas Baleares in Spain. They say that beauty is in the brain of the beholder and the brains of men and women work differently when appreciating art. They recorded brain activity when participants viewed a range of unfamiliar paintings and photographs of objects, landscapes and urban scenes. A region of the brain called the parietal lobe — it processes and interprets visual information about where objects are in space — was active in both men and women when they judged a painting or photograph 'beautiful'. While neurons on both sides of the brain were stimulated in women, only those in the right hemisphere were activated in men. It is generally accepted that the right hemisphere is associated with global visual attention and the left with local attention; that is, the right hemisphere sees the whole picture and the left the parts of the picture. The researchers suggest that perhaps women make use of both global and local features when 'admiring' art, while men rely only on global features. This finding makes it difficult to ignore a woman's judgment of beauty.

'Easy on the eye' is easy on the brain. The 'beauty-in-averageness' hypothesis says that people show preference for averages: prototypical or average-looking images of people are rated more beautiful or appealing than variations of the same thing. People have similar preference for average-looking dogs, birds, fish, car, even watches. Piotr Winkielman of the University of California says that prototypes are beautiful because you have seen them before and therefore easier for brain to process them. 'Beauty basically depends on what you have been exposed to and what is therefore easy on your mind,' he says. Beauty is after all only skin deep.

Seeing is believing, they say, but sometimes we believe before we see. Winkielman and Jamin Halberstadt of the University of Otago in New Zealand came to this conclusion after they asked volunteers to look at photographs of ambiguous facial expressions that had been labelled as either happy or angry. The volunteers then watched movies of the faces slowly changing expression from angry to happy. After the movie the volunteers were asked to identify the photos they had seen originally while the researchers studied the volunteers' facial movements. The researchers noticed that when viewing a facial expression they had once thought about as angry, volunteers expressed more anger themselves than did volunteers viewing the same face who had initially recognised it as happy. The result shows the way we initially interpret the emotions of another person favours our subsequent perception and reaction to their facial expression. This means two people could have different recollections about the same emotional face and both would be 'correct'. Remarks Winkielman: 'It's a paradox. The more we seek meaning in other emotions, the less accurate we are in remembering them.' Sometimes a poker face indeed helps.

How good are our 'educated guesses' when making a numerical judgment?

65

The $63,999 question

What's the freezing point of vodka? Well, it's not exactly a $64,000 question*, but would you have paid more attention to it if it was called a $63,999 question?

First, you are not expected to know the freezing point of vodka; you can arrive at an answer by adjusting the information you know. If you have paid any attention to science in your schooldays, you would know that alcohol freezes at a temperature lower than water. You can confidently make an educated guess that the freezing point of vodka is less than 0 °C or 32 °F, the freezing point of water. What you have done here is to anchor on information that comes to your mind and adjusting it until a plausible estimate is reached. Although this method of estimating by using reasoning and past experiences — better known as anchoring-and-adjustment heuristic — is often helpful, people's estimates are usually biased towards the initial value of the anchor, especially when a relevant numerical value is available.

Chris Janiszewski and Dan Uy of the University of Florida ran five experiments to test the idea how the precision of the anchor influences the amount of adjustment. The experiments used hypothetical scenarios in which graduate students were asked to make a variety of price estimates if they were buying, for example, a basket ball, a plasma TV or a beach house. The beach house scenario involved three different prices:

* A crucial or essential question concerning a particular problem or situation; the idiom comes from the title of a 1950s TV quiz show, *The $64,000 Question*.

one group of buyers was told a price of $800,000, another $799,800, and the third $800,200. When all the three groups were asked to estimate the 'real' price of the house, their average estimate came to be $751,867, $784,671 and 778,264 respectively. The group that started with a price tag of $80,000 — a rounded anchor — guessed much lower than the groups that started with a price under or over the mental anchor. That is, the group moved farther away from their anchor. Furthermore, most estimates of this group were in round numbers. That is, those who started with a round number as an anchor most likely guessed the 'real' price in round numbers.

The researchers ran this experiment with different scenarios and always got the same answer. To confirm their lab findings with the real estate sales, they compared list prices and actual sale prices in a county of Florida. They were not surprised when they found that sellers who listed their home more precisely — say $694,500 as opposed to $700,000 — consistently got closer to their asking prices. Their verdict: adjustment away from a numerical anchor is smaller if the anchor is *precise* than if it is rounded.

Translating this result to a retail scenario means that when, for example, buying a book priced $19.95 — a precise anchor — you might still think that it's priced too high but you're thinking in terms of cents and might think $19.25, $19.50 or 19.75 as a fair price. If the book is priced $20, your mind would be working in a different way thinking in terms of round numbers such as $18, $19 or $21. You know now why customers respond better to prices such as $ 1.95 or $1.99, just below the next dollar increment. Such a price gives them a precise mental anchor, a starting point for psychological manoeuvring to follow.

Janiszewski and Uy say that the anchor-precision effect may also be applicable in large number of real-world scenarios; influencing negotiations, assessment of health risk, and assessments of present and future value. Wray Herbert of the Association for Psychological Science in the US gives an example how it could be used to offer medical information in either precise or general terms: a doctor might say that your chance of responding to a medication is 'good' or that your chance of responding is 80 per cent. Wray, however, disagrees with Janiszewski and Uy's numerical precision, at least in this case: 'The percentage is more precise, but many studies have shown that patients prefer vague generalities like "good", so doctors tend to use them.'

66
The myth of the power of Mozart

Listening to the music of Wolfgang Amadeus Mozart may help you relax, but contrary to popular belief it doesn't make you smarter.

Mozart's musical sequences repeat themselves in a very logical and rhythmic way and even Einstein was entranced by Mozart's intensely patterned sonatas, at least when he described Mozart's music as 'so pure that it seemed to have been ever-present in the universe, waiting to be discovered by the master.' Though Einstein believed that music and reasoning ability were linked, listening to the master's pieces likes *Eine Kleine Nachtmusik* are not likely to make you an Einstein.

The 'Mozart myth' does not come from Einstein, but from an American study published in the international journal Nature in 1993. The study found that college students who listened to a selection of Mozart's 1781 *Sonata in D Major for Two Pianos* performed better in a reasoning test than those listening to some other 'relaxation music' or sitting in a silent room. The improvement in the reasoning ability was temporary and was gone in half hour.

The study results were widely reported in the popular media as evidence of what the media called 'the Mozart effect'. It was touted as a magic bullet to boost children's intelligence. A musician called Don Campbell, who was not a scientist and was not involved in the original research, was the first to see the commercial potential. He quickly trademarked the term 'Mozart effect' and without any evidence expanded the role of Mozart as the healer of mind and body. The rapid commercialisation —

streams of books and CDs — only amused the researchers who never made any claims that listening to Mozart made people smarter. Frances Rauscher, the lead researcher, commented in 2006, 'Nobody ever said listening to Mozart makes you smarter.'

Since 1993 many studies have tried to replicate the performance-enhancing effects of Mozart, but none has succeeded. The final curtain fell on the Mozart effect when in 2010 Vienna University psychologists gave a clear-cut verdict: there is no evidence that merely listening to Mozart's music can boost cognitive performance. Their finding was based on the analysis of 3000 cases from forty international studies.

The finding doesn't say that music has no impact on the brain development. Playing a musical instrument demands extensive procedural and motor learning that results in the plastic reorganisation of the brain. Changes take place in the motor and sensory cortexes that play an important role in motor skill learning. There is no confirmed evidence yet whether music lessons can make children smarter, but many studies have shown that they can indeed improve their verbal and spatial abilities (*next story*).

Listening to music has its own benefits: it depresses activity in the amygdala, the brain's region that registers fear and other negative emotions, which in turn can lower blood pressure and treat anxiety and insomnia. And it has been shown that playing classical and soothing music, not necessarily Mozart, can increase the milk yield of dairy cows.

For parents keen on playing Mozart for their babies to make them smarter, you may as well as be playing any music. Beyoncé, if you prefer.

Listening to music has emotional appeal, and learning music boosts brain

67

Music moves minds

You don't need an expert to tell you that music can stir the emotions. If you do need one, ask renowned neurologist Oliver Sacks who knows lots about music's emotional effect. 'Certainly music seems to be the most direct form of emotional communication,' he writes in his book, *Musicophilia: Tales of Music and the Brain*, 'It really seems to be as important a part of human life and communications as language and gesture.'

Humans have always created, listened and danced to music. Why has this 'heavenly art' captivated the entire human species? The answer lies in the human brain. If you look at the brain scan of a person while listening to music, it would appear the brain is on fire. Music simultaneously stimulates a remarkable number of regions on both sides of the brain, including regions responsible for emotion, perception, memory, motor control, time movements and language. 'Listening to music lights up, or activates, more of the brain than any other stimulus we know,' says Istvan Molnar-Szakacs, a neuroscientist at the University of California.

Mirror neurons are a type of neurons that respond equally when we do something and when we see someone else doing the same thing (*page 84*). There is growing evidence that emotional empathy ('I feel how you feel') also recruits mirror neurons in creating link between self and other. Molnar-Szakacs suggests that the mirror neuron system is linked with the limbic system, the so-called emotional hub of the brain. This link may explain why music has such a profound emotional effect. Once early humans invented music it quickly spread from culture to culture because of its emotional appeal.

Listening to your favourite music lights up reward and pleasure centres in the brain (the same centres light up when you eat chocolate or have sex), but learning to sing or play a musical instrument, particularly beginning at an early age, rewires your brain. While much research demonstrates that childhood music lessons improve brain functions, a new study now shows that those lessons could help boost your ageing brain decades later. 'Musical activity throughout life may serve as a challenging cognitive exercise, making your brain filter and more capable of accommodating the challenges of ageing,' says lead researcher Brenda Hanna-Pladdy of Emory University in the US. 'Since learning to play an instrument requires years of practice and learning, it may create alternate connections in the brain that could compensate for cognitive decline as we get older.' The study also suggests that duration of musical study during childhood was more important than whether musicians continued playing at an advanced age. Both the years of musical training and how early the training started are crucial for a larger impact on the brain.

Canadian researchers report the verbal intelligence of 4- to 6-year-olds rises after only one month of music education. The results were confirmed by brain scans before and after the music program. In the study, conducted at York University in Canada, two groups of children participated in a cognitive program in which animated cartoon characters delivered lessons on a screen. One group received music-based training that involved a combination of motor, perceptual and cognitive tasks and instruction on rhythm, pitch, melody, voice and basic musical concepts. The other group received visual art training that emphasised visuospatial skills relating to concepts such as shape, colour, line, dimension and perspective. Both groups received two one-hour training sessions each day, over four weeks. The researchers tested the children for verbal and spatial intelligence before and after the training. Results didn't show any significant increase in verbal intelligence of children who participated in visual art training program; they didn't either show any improvement in spatial abilities because visual-motor skills are developed later in life. However, 90 per cent of children who took part in the music-based training program showed remarkable improvement in verbal intelligence. The verbal IQ tests assess children's attention, word recall and ability to analyse information and solve problems using language-based reasoning. Other studies have found that children who receive musical training also have better reading skills. Music helps with language comprehension.

Catch them young. Though young brains are more malleable than adult ones, research shows you're never too old to reap benefits of musical training.

Losing weight seems to be
the ultimate mind-body problem

68

Memo dieters: this is your brain speaking

Our body weight depends on the balance between calories consumed and calories burned — a balance we can control by watching how much we eat and exercising regularly. Research on controlling body weight (body fat, actually) also agrees with this folk wisdom. Yet, generations of dieter tell the same sorry story: the lost kilograms (or pounds, if you prefer) soon come back, mostly with a vengeance. Decades of research has provided slim comfort for fat people except to come up with only one irrefutable fact: overeating makes you fat.

In evolutionary terms, out bodies' ability to store energy in the form of fat for future use was an absolute necessity for survival when food was scarce or unpredictable. The brain-hunger system that evolved over millions of years even erred on the side of gluttony by furnishing our brains with a system that made eating a highly pleasurable activity.

The system that regulates our eating behaviour is highly complex. It integrates information about the body's energy needs and the status of its fat stores, then initiatives changes in behaviour and energy processing in response. Specialised regions of brain stimulate feelings of hunger or satiety. According to *Scientific American* magazine, the brain receives three types of information: (1) stored energy status: leptin, a hormone secreted into the bloodstream by fat cells, indicating how much fat they contain; (2) metabolic status: circulating glucose represents energy immediately available to cells; and (3) neural and chemical signals from the gut indicating whether digestive organs are full of food. The brain then responds by changing body's food intake through appetite and satiety signals.

Ghrelin, a hormone produced by glands in stomach, signals the stomach's readiness for a meal to the brain. It's often called the hunger hormone. Leptin is called the satiety hormone as it suppresses appetite. Higher levels of ghrelin have been associated with obesity, fatness that compromises a person's health.

Many researchers have observed that each human body 'decides' what its maximum weight should be and adjusts its activities to maintain this weight — the set point — within a kilogram or two. They also claim that dieting cannot lower the set point: only physical exercise appears able to do so. The body experiences a diet as if it were faced with starvation and responds to conserve energy. Dieters unconsciously restrict their physical activity. When the dieting period is over the body continues to signal the need to eat until the fat stores are replenished. The reason for gaining the lost kilograms is not clear, and could include both behavioural and metabolic factors. Brain-imaging research, however, shows that successful dieters have increased activity in the brain areas involved in the control of behaviour.

If you love junk food, you can blame your brain. Glucose is the primary fuel source for the brain. A decline in blood glucose increases hunger sending signals to eat, particularly high-sugar and high-fat foods. A research team at Yale University in the US has found that obese people's brain are less able to control impulse to eat than people of normal weight, especially when their blood-glucose levels go down below normal. Their brain scans showed that the prefrontal cortex, the decision-making region of the brain, loses the ability to resist high-calorie foods such as fried chicken, hamburgers and pizza. The message to you is to eat healthy foods that maintain glucose levels as the brain needs it glucose.

Well, it's hard to resist fried chicken, hamburgers and pizza even if you are thin. Brain-imaging studies show that mere sight and smell of these foods release dopamine in the brain. The amount released depends upon our yearning to eat. If you badly want to eat, dopamine levels will shoot up. Don't let your desire go out of hand.

Now, the last bit of bad news for dieters. Dieters struggle to lose weight because a lack of nutrition forces their neurons to eat themselves, making the feeling of hunger even stronger. This claim comes from researchers from Yeshiva University in New York. Autophagy, meaning 'self-eating', is a kind of housekeeping process which appears in nearly every cell. Scientists previously believed that autophagy in neurons remained relatively constant, even during starvation.

If this news makes you crave for chocolates — it's not something you can control. Chocolate craving is probably induced by stress and governed by body chemistry. Studies show that this sweet, high-fat food fires up a network of brain regions — we won't list their long name as we don't want to keep you away from your chocolates any longer — and releases chemicals such as serotonin which make us feel relaxed.

Enjoy your chocolate — it's a good source of antioxidants — but make sure it's only a small piece (the darker the chocolate, better it is).

69

Eat right for a smarter brain

If you want your brain to be razor-sharp and resourceful, you need to eat right food. Science has yet to uncover fully the processes involved in the actions of food on the brain, but it's generally accepted that the relative abundance of specific nutrients in some foods can help in sharpening memory, boosting brainpower and counteracting the effects of ageing. Here's a brief look at the latest thinking on 'brain foods'...

So far, a diet rich in **omega-3** fatty acids (found in oily fish, flax seeds, chia seeds, walnut, kiwi fruit, pumpkin) has garnered most support from researchers. Its benefits include improving learning and memory and helping to fight against depression. 'Synapses in the brain connect neurons and provide critical functions; much learning and memory occurs at the synapses,' says Ferdinando Gomez-Pinilla of the University of California who has spent years studying the effects of food, exercise and sleep on the brain. 'Omega-3 fatty acids support synaptic plasticity and seem to positively affect the expression of several molecules related to learning and memory that are found on synapses.' Getting omega-3 supplements from food rather than from capsule supplements can be more beneficial, providing additional nutrients, he advises.

Curcumin (found in turmeric, a yellow spice used in Indian curries and American mustard) may prove helpful in the treatment of Alzheimer's disease (it has shown to reduce memory loss in animals with brain damage). It's probably no coincidence that high consumption of turmeric in India is related to the low prevalence of Alzheimer's disease (four times less than the US).

Supplements of some B vitamins have shown to have positive effects on memory. These vitamins are: B6 (various natural sources include meats, whole grains and starchy vegetables); B9 (also called folate or folic acid; spinach, orange juice, yeast and concentrated yeast extracts sold as Vegemite or Marmite are rich in folate); and B12 (not available from plant products). Deficiency of folate is associated with clinical depression in women.

Vitamin D (found in oily fish, mushrooms, milk, soy milk, cereal grains) is important for preserving cognition in the elderly.

Most likely you have heard about antioxidants. Oxidation in the body produces chemicals called free radicals. Antioxidants in foods neutralise these free radicals. The brain is highly susceptible to damage caused by free radicals. It consumes a lot of energy (about 20 per cent of our daily calorie intake) and the reactions that release this energy also produce free radicals. Good sources of antioxidants include: beta-carotene (found in pumpkin, mangoes, apricots, carrots, parsley); polyphenols (thyme and oregano); vitamin C (citrus fruit, blackcurrants, kiwi fruit, broccoli, spinach, capsicum and strawberries); and vitamin E (vegetable oils, nuts, green leafy vegetables, whole grains).

Flavonoids is a group of chemical compounds that has attracted much attention in recent years. Many studies have suggested that flavonoids may improve memory, learning, reasoning skills, decision-making, verbal comprehension and numerical ability. They have also shown to slow age-related decline in mental function. Scientists have so far identified more than 6000 flavonoids all of which are powerful antioxidants. Foods rich in flavonoids include cocoa, tea, berries, grapes, citrus fruits, spinach, peppers, onions, parsley, celery, dark chocolate, wine (higher in red wine) and soy foods such as tofu.

Coffee (not decaffeinated coffee) instructs the body to produce the hormone adrenaline (also called epinephrine). Known as the 'fight or flight' hormone, adrenaline creates a surge of glucose in the blood making you feel energetic. The flip side is that it also causes the body to release the hormone insulin, which makes you drowsy. A recent Harvard University study has found that women who drink two to three cups of coffee every day decrease their risk of becoming depressed by 15 per cent. The study involved 51,000 women whose coffee intake was monitored over a 10-year period. Similar results for men have reported by other studies. Beware drinking too much coffee can raise blood pressure and increase heart rate. Pregnant women are advised not to drink more than two cups a day.

Some studies indicate that moderate drinkers (two or less standard drinks per day) have less risk of Alzheimer's disease compared to teetotallers. Remember studies give alcohol a controversial verdict.

Foods to avoid: Saturated and trans fats adversely affect mental processes. Saturated fat is found in butter, ghee, lard, coconut oil, cottonseed oil, palm oil, meat and dairy products such as cream and cheese. Trans fat is found in fried food and many commercially baked products such as cakes, chips and crackers. Junk food and fast food are usually rich in these unhealthy fats.

Do not overeat. Excess calories can increase the vulnerability of brain cells to damage by causing the formation of free radicals. In fact, moderate calorie restriction below normal consumption levels could protect the brain by reducing damage caused by free radicals. In other words, occasionally skipping a meal or having a lighter meal is good for your brain.

In addition to eating right food, a healthy brain requires that you exercise regularly and have a good night's sleep.

'Left brain' and 'right brain'
people differ is indeed a myth

70

I'm left brained and that's why ...

You have probably heard that left-brained people are logical and good at mathematics and right-brained people are artistic and bad at mathematics. The widely held belief that we use the left brain for logical, analytical thinking and the right brain for intuitive, artistic thinking is not simply true.

This myth can be traced back to the days of nineteenth-century craze of phrenology (even Herman Melville made his Moby Dick narrator, Ishmael, an amateur phrenologist). Phrenologists studied shape and size of the head to determine a person's character and mental abilities. They believed that different mental functions were located in different organs of the brain, and the growth of the various organs was related to the development of associated mental abilities. As this growth would be reflected in the shape of the skull, personality traits could be determined by reading bumps and depressions on the skull. For example, if you move your finger on the back of your neck, you will notice a bump formed by the base of your skull. This bump, according to phrenologists, defined the attachment of sexes to each other.

In the twenty-first century this bump-reading nonsense has been replaced by sophisticated brain-imaging techniques. Research reveals less romantic sides of the brain: there is no evidence that the left brain is 'mathematical' and the right brain 'musical'. Yes, the brain is divided into two hemispheres. They look almost identical anatomically, but they are not independent. They are connected by thick bundles of nerve cells which carry information from one side to the other. The two hemispheres differ not so much in what they do, but in how they process tasks. The left hemisphere is better at details (such as recognising a particular face in a crowd), whereas the right hemisphere is better at dealing with a general sense of space (the relative positions of people in a crowd). In the case of language, for example, the left hemisphere focuses at step-by-step processes, such as grammar and word generation, whereas the right hemisphere focuses at feeling a rhythm, such as intonation and emphasis of speech.

There are no specific 'left brain' or 'right brain' cognitive functions. Both hemispheres work in concert with each other, whether we are reading, painting or solving an algebra equation. The brain is remarkably adaptive and children who have had hemispherectomy — in which half of the brain is removed — the remaining half overtakes most of the functions of the missing half (*page 206*). Whether you are young or old, learning actually strengthens and creates new connections between neurons.

Don't bother to take internet quizzes to find out whether you're a 'left brain' or a 'right brain' person. Like Chinese Yin Yang symbols the two hemispheres of your brain are in perfect harmony. Use your whole brain if New Age hucksters try to sell you a book or program to boost capabilities of your left or right brain.

71

Relax to generate alpha waves

Electroencephalograph (EEG) is one of the first tools used to record brain activity. An EEG records electrical impulses, or waves, generated by the brain by hooking up electrodes to the head.

Hans Berger, a psychiatrist at the University of Jena in Germany, was the first person to record electrical activity in the brain. In the early 1920s he realised that the best way to study the human brain was neither by dissection nor psychoanalysis, the two methods known at that time, but by recording its electrical activity. He thought if the heart produces electricity activity (which can be recorded as an electrocardiogram), then the brain must also produce electrical signals. Using a galvanometer, a device used to record electric current, he started experimenting on his patients. His early experiments were on patients who'd lost some of their skull bones in surgery. He made his first recordings in 1924 when he placed electrodes under the scalp of one of these patients. Later he succeeded in getting recordings from healthy people, including his teenage son.

A shy and aloof man, he worked in utmost secrecy for five years before he published his results. He reported that the brain generates rhythmic electrical impulses or 'brainwaves'. He identified two types of waves: alpha waves and beta waves. The waves change dramatically if the subject simply shifts from sitting quietly with eyes shut (alpha) to fully alert (beta). He called his recordings electroencephalographs (Greek for 'the writing of the brain').

Initially, Berger's discovery was completely ignored in Germany. When he was introduced as the most distinguished of all the visitors at an international symposium in Paris, tears welled in his eyes as he said: 'In Germany I'm not so famous.' He died a lonely and dejected man. His EEG is still a widely used technique in neurology. Its major drawback is that like other brain-scanning techniques it cannot show activity in the specific regions of the brain. The procedure is very safe and easily provides the record of how the brain functions over time. It's used to evaluate head injuries, tumours, sleep disorders, Alzheimer's disease and other brain disorders.

We know now a little more about brainwaves. Here're the five waves that make up the graphical record of brain activity produced by the EEG machine (the frequency range shown below are approximate):

Delta Waves (frequency up to 4 hertz or cycles per second) are associated with unconscious mind, dreamless deep sleep.

Theta waves (frequency 4 to 7 hertz) are present during light sleep, in subconscious mind, in minds of long-term meditators, and inhibition where a person is actively trying to repress a response to action. They are more common in children than adults.

Alpha waves (8 to 13 hertz) are associated with relaxed mind or daydreaming; they are easily produced when quietly sitting in relaxed position with eyes closed.

Beta waves (14 to 30 hertz) are the most common form of brainwaves and are associated with fully awake mind or normal thinking state.

Gamma waves (above 30 hertz) occur during sudden sensory stimuli — learning, memory formation and perception.

The reason for sleep remains one of
the greatest mysteries of neuroscience

72

The science of counting sheep

Sleeping is an instinctive behaviour — a natural impulse. Ask anyone why we have the sleep instinct and they would instinctively reply that we sleep to refresh the body and the mind.

Scientists have not yet identified any vital biological function that sleep restores; however, most physiological functions — for example, heart rate, respiratory rare and metabolism — are different in sleep. It's the mind — not the body — that benefits most from sleep. But not much is known about how sleep benefits the mind. It is known that sleep allows neurons to strengthen or weaken connections with other neurons. This neural remodelling helps in consolidating the streams of information gathered while we are awake. This remodelling, in turn, helps in memory and learning by repackaging memories for long-term storage in the brain. It also helps create associations between unrelated ideas, enhancing creativity. Research shows that if the processing of information during sleep is disturbed by insomnia and other chronic sleep disorders it could lead to psychological disorders, including depression, attention-deficit disorder and attention-deficit hyperactivity disorder. Research also suggests that sleep helps in fat reduction.

Sleep well, to keep yourself sane — and slim.

We spend nearly one-thirds of our lives asleep. We are not alone; all animals sleep in one form or another. Body size appears to be related to the amount of sleep a species need. The larger the animal, less sleep it needs: elephant (about 3 hours), humans

(8 hours), dog (10 hours) and cat (12 hours), platypus (14 hours). The reason perhaps is that neurons in the brains of smaller animals are more prone to injury because of higher metabolic rate. Consequently, their neurons require more time for strengthening connections with their partners.

Studies of the brain during sleep show that it is highly active while the body is passive. The active brain circuits produce waves which can be recorded by an EEG (*previous story*). The recordings — and the measurement of the eyes and the limbs — show five different stages of sleep:

Awake: Sleep-on neurons, a small group of neurons in the forebrain responsible for inducing sleep, are inactive; alpha brainwaves (relaxation)

Stage 1: Marks the transition between awake and asleep; sleep-on neurons fire; shallow brainwaves; muscles relax and eye movement slows down

Stage 2: It lasts the longest; bursts of wave activity; sleep talking

Stages 3 and 4: It last for about 30 minutes; deepest, or slow-wave, sleep; delta waves appear; sleepwalking and bedwetting

Stage 5 or rapid eye movement (REM) sleep: Lasts 10 to 15 minutes; accompanied by rapid, jerky eye movements; heart rate, blood pressure and body temperature becomes much more variable; the brain is highly active (it's on fire); vivid dreams occur (dreams also occur in other stages but they are not vivid)

During the night, these stages of slow-wave and REM sleep are repeated in roughly 90-minutes cycles until waking occurs. There is more slow-wave sleep early on and more REM sleep towards morning.

Free radicals damage neurons when they are active. During non-REM stages of sleep certain neuron receptors become inactive and this inactivity may allow neurons to repair membranes damaged by free radicals. REM sleep time is highest early in life and falls as we grow older. It is believed intense neural activity during REM sleep may allow the brain to develop properly. 'REM sleep, however, is the proverbial riddle wrapped in a mystery inside an enigma,' says Jerome M. Siegel of the University of California. 'The cell-repair hypothesis could explain non-REM sleep, but it fails to account for REM sleep.'

Don't lose sleep over this scientific matter. Back to counting sheep.

73

Dreams, dreams, dreams

Most dreams, but not all, occur during the rapid eye movement (REM) period of sleep. We spend nearly one-thirds of our lives asleep, and one-fifth of this time is REM sleep. This means we spend one-fifteenth — that's a big slice — of our lives dreaming. Why?

When we think of dreams, we think of Sigmund Freud. He didn't answer this question as REM sleep was not known in his time (it was discovered in 1953). His efforts were focused on interpreting dreams: suppressed desires rise up in our dreams disguised as symbols which therapists can usefully decode. Dream research has now left therapists' couch.

We spend so much time dreaming there must be a biological relevant reason for this. Some of the reasons proposed by researchers:

- Dreaming helps in consolidating information (*previous story*).
- 'We dream to forget'. This 'reverse learning' hypothesis — it comes from Francis Crick (the co-discoverer of the double helix structure of DNA) — allows the brain to discard useless or less important information collected during the day.
- Dreaming leads to creativity (*page 14*).

Do we actually 'see' dreams? According to psychologist Robert O. Duncan of the City University of New York, there are two competing views. Studies show similar patterns of activity in the visual cortex, the largest brain area devoted to vision, during daytime vision and dreaming. Other view is that dreams originate in the region that store memories and then connect to visual brain areas. 'This theory would explain why dream images are only as detailed as our memories,' he says.

This theory also demystifies the myth that our dreams happen in shades of grey rather than in colour. They are as colourful and vivid as your memories. Of course, this theory doesn't explain another myth that if you die in dream, you really die. It's not true. However, it's true that dreams change amazingly little with age. In general, there is more aggression in dreams than friendliness, says C. William Domhoff of the University of California who likes to analyse personal dream journals.

Don't worry about aggression in your dreams. Memories of dreams usually disappear soon after waking up. Anyway, 'we dream to forget' as Francis Crick said. Our brains are damn good at forgetting the information it considers useless. Once memory storages in computers and smart phones become as smart as our brains, we won't have to think about erasing memory to create space for the new stuff we want to download.

74

You're doing a terrific job

Everyone needs feedback. It helps us in making informed decisions. Here's what research says how anyone responsible for evaluating the performance of others can tailor their feedback to make it more effective.

Eliminate the negative. Our brains have 'negativity bias': they're simply built with a greater sensitivity to bad news. Research by social psychologist John Cacioppo of the University of Chicago has shown that electrical activity in the brain surges higher in response to stimuli it deems negative. The bias is so automatic that it can be detected at the earliest stage of the brain's information processing. As negative news weighs more heavily on the brain, negative feedback is likely to hurt most.

'Bad is stronger than good,' says organisational psychologist Robert Sutton at Stanford University. 'It's more important to eliminate the negative than accentuate the positive.' Consequently, he advises, negative feedback is very, very difficult to do well. Don't criticise unless you have been asked to do so. Unsolicited criticism can only bring frustration.

However, if you're giving feedback to neurotic individuals, listen to Jacob Hirsch and Michael Inzlicht, both at the University of Toronto: Neurotic individuals 'prefer the devil they know over the devil they don't know'. They would rather receive clear negative feedback than uncertain feedback, even though the outcome of the uncertain feedback could potentially be positive. 'Uncertainty can be very stressful,' the researchers say, 'and high levels of neuroticism contribute to this dislike of the unknown.' The researchers noted clear stronger activity in the anterior cingulated cortex, a region of the brain associated with conflict-related anxiety, of neurotic individuals when they were given uncertain feedback, compared to when they were given unambiguous negative feedback.

Make it timely. A study by psychologists Keri Kettle and Gerald Häubl of the University of Alberta reveals an interesting relationship between the timing of feedback on grades and its influence on performance. For their experiment they recruited 271 college students by email which was sent 1, 8 or 15 days before test of their performance on a 4-minute oral presentation. Each student was given an anticipated feedback time, which randomly ranged from 0 (same day) to 17 days. Students were also asked to predict their performance. The presentations were rated by their classmates on a scale from 0 (poor) to 10 (excellent). The study results show that students who were told they would receive feedback quickly earned higher grades than students who expected feedback at a later time. Furthermore, when students expected to receive their grades quickly, they predicted that their performance would be worse than students who were to receive feedback later.

Mere anticipation of more rapid feedback improves performance because the threat of disappointment is more prominent. 'People do best precisely when their predictions about their own performance are least optimistic,' the researchers say.

Discuss it. Social scientist Marieke van der Schaaf of Utrecht University in the Netherlands advocates the idea of students maintaining digital development portfolios in which they collect their own work and their reflections on how and what they have learned. If teachers talk to students about their portfolios, her research shows, it supports students in their self-reflection process. 'Providing supportive feedback not only means that the teacher informs the students about what is good or less good but also why that is the case and how better results could be achieved,' she says. 'The students should preferably be given the opportunity to respond to feedback so that a dialogue develops in which the student can pose to the teacher and the teacher can check if the students has understood feedback.'

Put simply, verbal feedback encourages thinking.

75

Craving for your next internet fix

'I knew I needed help when I found myself tapping my e-mail "refresh" button like a lab rat trying to get cocaine,' confides Katherine Ellison, a Pulitzer Prize-winning journalist, in a column in *Los Angeles Times.*

Like Ellison, do you feel anxious that you have not read your email or have checked your Facebook or Twitter feeds for some time? Do you spend increasing amounts of time online and ignore your family, friends and work? Do your family and friends complain to you about the amount of time you spend online? Do you lie to them about how much time you spend online? Do you become defensive or secretive when someone asks you what you do online? Do you become anxious or irritable when you can't go online?

If your internet use interferes with your daily life, you're probably addicted to it. The American Psychiatric Association describes internet addiction as a compulsive-impulsive disorder with three subtypes: excessive gaming, cyber sex and e-mail/texting. The Association provides four criteria: (1) excessive use, often associated with loss of time; (2) withdrawal, including feeling of anger, tension, depression when the internet is inaccessible; (3) tolerance, including the need for better hardware and software; and (4) negative repercussions, including arguments, lying, social isolation and fatigue.

Some psychologists say that internet addiction is a problem as difficult to beat as being hooked on drugs, alcohol or gambling. Others say it's not true addiction. Problematic internet use is no more than a symptom of other existing disorders such as anxiety and depression which need treatment. So-called internet addicts use the internet as coping mechanism against underlying psychological problems. It's not the technology which is addictive, it's the behaviour.

Chinese, however, take the problem of internet addiction more seriously. As many as 24 million Chinese urban youths can be labelled as internet addicts. The number may seem extremely high, but not when you put it in the context of Chinese culture. Children face extreme pressure from parents to perform (probably the kind of parental pressure described by Ami Chua in her book, *The Battle Hymn of the Tiger Mother*). People work very hard, up to twelve hours a day, six days a week. There are plenty of cheap internet cafes around the corner on most blocks. 'Sometimes the internet is the greatest and only escape,' says Karen M. von Deneen of Xidian University in China. 'In online games you become a hero, build empires and submerge yourself in a fantasy. That kind of escapism is what draws young people.'

China not only has the highest number of internet addicts in the world; it's also at the forefront of research on internet addiction. A team of neuroscientists at Xidian University hints that excessive internet use can physically rewire the brain. Their brain scans of adolescents with long-term internet addiction showed a reduction in grey matter as well as white matter in several areas as much as by 20 per cent. Significantly, the longer the addiction's duration, the more distinct was the loss. Grey matter, which consists of closely packed neurons, makes up the cerebral cortex, the outer layer of the brain. White matter, on the other hand, is deep in the brain. It mainly consists of nerve fibres called axons which link together various regions of the brain. The shrinkage of brain tissues could lead to short-term memory damage, harm decision-making abilities, reduce inhibition of inappropriate behaviour and diminish goal-orientation, the neuroscientists warn.

If you're craving for your next internet fix, try some relaxation techniques such as deep breathing or meditation, socialise with your family and friends, or follow this version of an old proverb: after dinner, rest a while, after a long internet session, walk a mile.

Spoiler alert: Our brains like stories — and spoiling the ending doesn't spoil our enjoyment

76

Tell me a story about my brain, but first tell me how will it end

American neuroscientist Micahel Gazzainga is famous for his study of 'split-brain' patients (whose brain hemispheres have been surgically disconnected) which revealed the brain's split personality: its left and right hemispheres are specialised in different tasks. For example, the left side is better at analytical and verbal tasks, while the right side deals with a general sense of space. Gazzainga calls the left hemisphere 'the interpreter' as it takes what information it has and turns it into a meaningful tale. It happens all the time; for example, you overhear a fragment of gossip and you fill the gaps with assumptions. The left side of the brain has the storytelling 'machinery' which turns our feelings, actions and experiences into a running narrative. This is what creates our sense of a unified self, says Gazzainga.

When a storyteller and a listener 'unify' — the two really understand each other — the activity in their brains looks remarkably similar. Princeton University researchers noticed this coordinated brain activity, or 'neural coupling' as they call it, when they recorded brain activity of participants listening to a woman recounting an amusing story. The stronger the match between the brains of the speaker and the listener better the listener understands the story. Listening to stories connects to you others, indeed.

A brain-imaging study by Jeffrey Zacks of Washington University and his colleagues showed reading stories is not a passive activity: we are so much influenced by them that we adopt the role of the protagonist as our own. The researchers tracked the brain activity of participants as they read short stories, finding that the same regions fire up as if they were in real-life situation of the character they were witnessing. For example, if the character in the 1951 book *One Boy's Days* used by researchers 'picked up a pencil', brain activity increased in the frontal lobe which controls grasping motions. The study suggests that readers do mental simulation when reading a story.

These days it seems no story on the brain is complete without mentioning the hormone oxytocin (*pages 185 and 209*), variously described as a 'compassionate hormone', 'trust hormone , 'cuddle hormone, 'love molecule' and even 'moral molecule'. Paul Zak, a US neuroeconomist at the forefront of research on oxytocin, believes it's oxytocin that makes us empathise with fictional characters. When volunteers in one of his experiments watched a 5-minute video telling the story of a 4-year-old boy with terminal cancer, their oxytocin levels increased on average by 47 per cent compared with volunteers who saw an emotionally neutral film about the same boy going to the zoo.

William Casebeer of the US Defense Advanced Research Projects Agency advances this idea further when he says, 'If I were a betting man or woman, I would say that certain types of stories might be addictive and, neurobiological speaking, not that different from taking a tiny hit of cocaine.' Using fMRI he has studied brain's reward and pleasure centres when people listen to a story. The centres light up the same way when we have pleasurable experiences such eating chocolate or having sex.

Now twist in this 'addictive' tale. If you are one of those nosy types who read the last few pages of a mystery novel first to find out how the story will end, relax as you have done the right thing. You are likely to enjoy the novel more than those who think spoiler will spoil their enjoyment. This assurance comes from psychologists Nicholas Christenfeld and Jonathan Leavitt of the University of California. They say that contrary to popular belief, knowing the ending of a story enhances readers' enjoyment of the story. They recorded readers' reaction to spoiled and unspoiled versions of twelve classic short stories, by authors including Agatha Christie, Anton Chekhov and John Updike. The readers strongly preferred the spoiled stories, particularly when an ironic twist appeared in the story. 'Once you know how it turns out, you're more comfortable processing the information — and can focus on a deeper understanding of the story,' says Leavitt. Perhaps it explains why a favourite book could be read many times without undiminished pleasure.

The hormone oxytocin is believed
to plays a role in developing trust

77
The smell of trust

Build a little fence of trust
Around today
Mary Frances Butts, 'Trust' (1919)

If research on the role of the hormone oxytocin in developing trust and social attachment is right, the only thing you have to do to 'build a little fence of trust around' you is to sniff oxytocin (yes, synthetic oxytocin is available as a nasal spray).

Take a sniff and you're sure to trust US neuroeconomist Paul Jack's trust in oxytocin, the hormone released during childbirth, breastfeeding and orgasm. To Jack, who is credited with the first published use of the term 'neuroeconomics', oxytocin is the 'social glue' that binds families, communities and societies and fosters trust between strangers. He and his colleagues have devised a 'trust game' to show how oxytocin boosts an individual's natural inclination to trust a stranger.

In the basic trust game, two players, who have no face-to-face contact, are told the rule in advance and are promised a fee, say $10, which is recorded in a computer account. When the game starts, player 1 can transfer some, all or no money to player 2. If money is sent, it's tripled and credited to player 2's account. If, for example, $6 is given player 2 ends up with $28. Player 2 can choose to return some or nothing to player 1. Player 1's trust is measured by the amount sent to player 2, while player 2's trust is measured by amount returned to player 1.

What happens after the players inhale an oxytocin nasal spray? The results show: (a) player 1 is sent 17 per cent more money than control players who took a placebo did;

(b) twice as many player 1s (almost half the total) who received oxytocins gave all their cash to their partners; and (c) players 2s showing the highest oxytocins levels returned the most money to their partners. From this and many other rigorous studies Zak and his colleagues' conclusion is that oxytocins increases trustworthiness.

What's the use of this finding in real-life situations? Do we spray oxytocin on each other's noses to improve social connectedness and trustworthiness? No, but it may help people with a social phobia or autism. Autistic people may have oxytocin dysfunction. It is very likely that oxytocin could help.

Research also shows that oxytocin is not the warm and fuzzy 'elixir' some people spruik it to be. A series of experiments by Carsten de Dreu of the University of Amsterdam has revealed that the effects of sniffing oxytocin are more nuanced than previously thought. In one experiment Dutch students were tested on a series of famous moral dilemmas (for example, a runaway trolley is rolling down the track towards five workmen who will die unless you pull a lever to divert it into the path of another lone workman, *page 208*). In all the dilemmas, the lone person had either a Dutch, German or Arab name, while the persons in the group were nameless. After a sniff of placebo, the participants were just as likely to sacrifice the lone person, no matter what name they had. But after sniffing oxytocin, participants were far more likely to sacrifice the lone person with either a German or an Arab name than the Dutch name. In all the experiments oxytocin made participants more positive towards people who belonged to the same group.

After a review of the scientific researches on oxytocin, Andrew Kemp and Adam Guastella of the University of Sydney have linked oxytocin to gloating, envy and aggression. Rather than supporting all social emotions, they think, it may increase the desire to approach people in social situations. Psychologists call it approach-related emotions. 'While approach-related behaviours are generally positive (such as trust and empathy), they also include negative emotions such as anger and aggression,' they say.

We cannot spoil the ending of the story (*previous page*) of oxytocin as the story has not yet ended.

78

Bytes in the brain

Talk about the brain to computer geeks and the first thing they want to know is the memory capacity of their brains — in bytes, the unit of digital information. The question is difficult to answer. It can be answered only if we assume that the brain is like a computer running software and therefore it has certain logical and computational abilities (in this context the term 'wetware' is used to describe the brain).

The brain has about 100 billion neurons and each neuron connects to other neurons. The estimates of the number of connections — the contact points are called synapses — varies from 1,000 to 10,000. If we assume each synapse could hold only 1 bit — eight bits equal one byte — of information and it makes 10,000 connections, the storage capacity could be 125 terabytes (or a thousand gigabytes). If we assume it could hold more information, say 8 bits or 1 byte, the memory capacity could be 1 petrabyte (or a million gigabytes) or more.

'Yet neurons combine so that each one helps with many memories at a time, exponentially increasing the brain's memory capacity to something close to around 2.5 petrabytes,' estimates US psychologist Paul Reber. However, he warns that the brain's exact storage capacity for memories is difficult to calculate. 'First, we do not know how to measure the size of a memory,' he writes in *Scientific American Mind* magazine. 'Second, certain memories involve more details and thus take up more space; other memories are forgotten and thus free up space.'

UK neuroscientist Chris Frith looks at the problem from a different angle. 'The brain has many limitations, but storage capacity is not one,' he writes in *New Scientist* magazine. 'The problem is getting the stuff in and, even more problematic, getting the stuff out again ... vivid images are necessary to get the information into our brain and get it out again later.' Vivid images eat up lots of bytes, we know.

Recent research suggests that given the right conditions, the brain can record an amazing amount of information — its memory capacity is much bigger than previously thought. Research also suggests that a single neuron can hold a memory. It's impossible to estimate the size of a neuron's storage capacity in terms of bytes. Neuroscientists have other things to worry about than bits and bytes.

If you're a geek, forget about comparing your brain to a computer. Your brain is far smarter than any computer you will ever see in your lifetime. And don't worry, it will never run out of memory. Our brains are already as smart as a neuron-based brain can be (*page 238*).

Most headaches do not arise from the brain;
however, they do activate pain centres in the brain

79

'I woke up with a freight train in my head'

Count yourself lucky if you don't have to use such metaphors to describe your headaches, as does the hero of Scott Wentworth's *Gunmetal Blues*, a 1992 musical mystery spoof of 1940s hard-boiled detective thrillers. Count yourself double lucky if you never suffer from an occasional headache: up to one adult in twenty-five has headache nearly every day. A recent survey by Eurolight, a European project to highlight the impact of headache, has estimated that 61 per cent women, 45 per cent men and 53 per cent children and youth suffered a headache during the past twelve months; and the chance of suffering a headache over a person's lifetime is 77 per cent.

Most headaches are not associated with any underlying problem in the brain. They are called benign or primary headaches. If your headache disappears with over-the-counter pain relievers you do not need to seek further remedy. Secondary headaches, such as tensions headaches and migraines, appear as a symptom of an underlying disorder. If these headaches become more frequent or severe, you need to see your doctor. Any severe headache occurring suddenly for the first time, especially if accompanied by neck stiffness, requires immediate medical attention.

The brain has a pain-control system that damps down the perception of unwanted stimuli. The brain region known as the anterior cingulated cortex is the key regulator of pain signals: it lights up when we feel pain, and also when we are emotional. Other regions of the brain involved in pain perception are the prefrontal cortex, the 'executive' region, and the amygdala, the emotion hub. For example, when we have a headache from excessive contraction of neck muscles, the brain's pain centres are activated. They now produce neurotransmitters such as serotonin and noradrenaline (also called norepinephrine) which pass on the pain-control message from pain centres to cells in the central grey region of the midbrain. This region then relays information downstream to the cells in the brain stem, the lower part of the brain joining the spinal cord. Activation of this control system dampens pain. It is likely that in people prone to headaches this pain control system is more delicate.

Ice-cream headaches, which are more common in people susceptible to migraines, occur when we quickly eat or drink very cold food. This sudden pain in the forehead can be easily avoided by eating cold foods more slowly.

Sometimes alcohol can cause headache by diluting blood vessels in the head. 'Morning after' hangover headaches are often caused by the breakdown of alcohol into acetaldehyde and acetate, which cause painful relaxation of arteries in the skull. You have probably heard of numerous hangover cures. Whatever you try it's not likely to work.

Air travel headaches occur during the descent if your nose is blocked. Use a nasal decongestant or block ears with cotton wool or ear plugs. This will reduce the effect of changing air pressure in sinuses and ears.

Mobile phone headaches are phantom headaches. They don't exist. According to a World Health Organization fact sheet (June 2011), 'To date, no adverse health effects have been established as being caused by mobile phone use ... At the frequencies used by mobile phones, most of the energy is absorbed by the skin and other superficial tissues, resulting in negligible temperature rise in the brain or any other organs of the body.'

Neuromarketing: it may have
a sexy name, but does it work?

80

They want to scan consumers' brains

Neuromarketing — using brain-imaging techniques to examine the brain's response to products and brands — attracted worldwide media attention in 2003 when highly respected American neuroscientist Read Montague decided to find an answer to Coke-Pepsi conundrum: why more people buy Coke than Pepsi while both drinks have similar taste. In his landmark experiment, Montague asked participants to drink either Coke or Pepsi without telling them what they were drinking while he watched their brain activity with an fMRI machine. The participants were evenly divided in their preference for the two brands. When the participants were told which samples were Coke, three-fourths said Coke tasted better. At the same time their prefrontal cortex, the brain's decision-making centre, lit up brightly. Montague concluded that images and ideas from Coke commercials were overriding the taste buds.

Here was the 'scientific proof' that marketing could influence consumers' brain. Realising the potential of marrying neuroscience with marketing, some marketing types immediately started styling themselves as neuromarketing gurus. They ditched their surveys and focus groups in favour of expensive but impressive fMRI machines.

In 2007 the new industry got a shot in the arm when Stanford University neuroscientist Brian Knutson and his colleagues used fMRI to scan participants' brain activity as they made purchasing decisions by evaluating products and prices on computer screens. The researchers found the activity in brain regions associated with anticipating gain correlated with product preference, while activity in brain regions associated with anticipating loss correlated with excessive prices. Their verdict: activity of distinct brain regions related to anticipation of gain and loss precedes and supports consumers' purchasing decision; and therefore can be used to predict purchasing decisions.

Neuromarketing experts, who like to keep their neuromarketing efforts secret, claim that it provides knowledge about customer preferences and what makes promotional campaigns more effective. However, some neuroscientists say that brain scans do not necessarily provide objective evidence. Brain regions do many things, not just one; and it's impossible to say whether increased or decreased activity in a particular brain region is 'better' or 'abnormal'. Even if brain scans correctly show that certain products fire up brain's reward centres, it doesn't mean it leads people to buy more products.

Martin Lindstrom, a US branding expert and author of *Brainwashed: Tricks Companies Use to Manipulate Our Minds and Persuade Us to Buy*, relates an experiment in which he looked at participants' brain activity as they viewed consumer images involving some major brands and religious images like rosary beads and a photo of the Pope. He found that the brain activity was uncanny similar when viewing both types of imagery.

Self-help books promise to improve every aspect of your life, but do they really help?

81

Oracle or snake oil?

'The greatest results in life are usually attained by simple means, and the exercise of ordinary qualities.' This advice comes from the first self-help book published in English: *Self-Help; with Illustrations of Character and Conduct* (London, 1859). Within a few years it sold more than 250,000 copies and made its author Samuel Smiles a celebrity. The book is still in print and has a respectable Amazon Best Seller Rank (free e-book is available from Project Gutenberg).

Self-books are now a multi-billion-dollar-a-year business worldwide. Check physical or virtual shelves of a major bookstore and you will find a self-help book on every conceivable aspect of your life, which offer innumerable mantras, remedies and programs. Do they work?

Hal Arkowitz and Scott O. Lilienfeld are American psychologists who share interest in helping the general public to distinguish myth from reality in the field of mental health. They see raising false hopes as one of the major problems with self-help books. 'Some self-help books may be unable to deliver on their expansive promises,' they say. 'As a result, readers may perceive their lack of change as personal failures and even see themselves as hopeless cases ("false hope syndrome"). When unreasonable expectations for self-change go unmet, people feel frustrated and despondent and may give up trying to change.'

This doesn't mean, in case of mental health, all self-help treatments are false. Some do have substantial medical evidence on their behalf, especially for treating some forms of anxiety, depression and trauma. But the advice to follow a self-help treatment should come from a health professional, not from a self-help book you have picked up from a bookstore.

Self-help books often make overblown claims which are not based on any scientific evidence. Furthermore, these books can steer a reader in the wrong direction and avoiding treatments that actually work. In the case of inspirational self-help books about coping with loss or divorce, the message also comes with a risk. If the reader is unable to cope, it may lead to frustration. Sometimes, readers start blaming themselves for their failures, for things that are out of their control.

Yet, the success of self-help books is not going to fade away soon as they have something going for them. They give people the opportunity to 'treat' themselves without seeing a doctor, psychologist or counsellor. They also make them part of an in-group. Someone has said that there is no shame in finding your G-spot along with everyone else.

Then there is the placebo effect. Self-help books act like placebos. The belief that you can improve is a powerful factor. Perhaps this is the reason why alternative therapies work; believers have great faith in their therapy and are able to convince their patients – readers, in case of self-help books — of the effectiveness of the treatment. You're healing yourself, with a little help form a self-help book.

'Buyers beware' rule also applies to buying self-help books. Think before you buy.

82

Worry rusts the brain

Worrying — random negative thoughts about a future event related to money, relationship, career or some other aspect of life — is part of our everyday lives; and still we go one with our lives without any appreciable effects on our health. Women worry more than men. This is not merely a matter of personality; this gender bias has been located in neurons in a specific region of the brain.

A little worrying is not a bad thing: it keeps us on our feet to take action and resolve problems. But when we can't get negative thoughts out of our head, it becomes a problem — chronic worrying. It's chronic worrying that's worrying. If excessive, constant and realistic worry continues for months or years and starts interfering with daily life, it becomes a mental disorder — anxiety. If we have irrational, recurring thoughts, say about germs and we wash our hands several times in an hour because we fear they are contaminated, worrying becomes obsessive and it acquires a new name, a medical term — obsessive-compulsive disorder.

'Worrying uses up mental resources (working memory capacity), particularly in people prone to worry,' explains Eleanor Leigh of King's College London. 'This means that there are fewer resources available to push away these unpleasant worrisome thoughts or think about something else. This can leave people caught in a cycle of worry.'

To break this 'cycle of worry' before it puts you under pressure (*next story*), try to:

Change the way you think. People who worry tend to see information in a negative way even if the information could be seen in a positive way, whereas people who do not worry tend to see the same information in a positive way. Get rid of negative information immediately.

Accept uncertainty. Change is unpredictable and induces worry, but it should not make you to lose control of your life. Once you accept the fact that you live in a changing world and humans have been coping with change since the dawn of history, it will become part of your nature to face the challenges of change without worrying about them.

Reframe your worry. This advice comes from US psychiatrist Robert A. Leahy, author of *The Worry Cure: Seven Steps to Stop Worry from Stopping You*. 'What happens if a worry comes true? Could you survive losing your job or being dumped? Reframing how you evaluate disappointments in life can take the sting out of failure.'

Once you have stop worrying, you can help others who experience excessive worry — anxiety disorder — which makes it difficult to complete important tasks. It also causes health problems such as insomnia, muscle tension and a variety of aches and pains. Unless chronic anxiety is treated it brings on major depression. To help someone dealing with anxiety (with the guidance of a mental health professional), Anxiety Disorders Association of America, advises friends and family: (a) to recognise and praise small accomplishments; (b) to modify expectations during periods of anxiety; and (c) try to be flexible and give needed 'space'.

83
Under pressure

Stress is both good and bad. One of the things it does is to promote the secretion of hormones adrenaline (also called epinephrine) and cortisol from the adrenal glands. These hormones prepare the body in a heightened state of alert — 'fight or flight'. Adrenaline mobilises energy and delivers it to muscles. Cortisol promotes energy replenishment and increases cardiovascular function so oxygen can travel more quickly to cells in the body.

The two hormones re-establish or maintain body's homeostasis — that is, various measures such as temperature and glucose level are as close to 'ideal' as possible — which had been knocked out of balance by stress. Over the short run, the two hormones provide extra energy to cope with stress. But if stress continues for prolonged periods, the continued secretion of adrenaline and cortisol becomes toxic. Long-term overexposure to these hormones could result in disorders such as high blood pressure, clogged arteries, erectile dysfunction and disrupting of menstrual cycle.

Hippocampus, the brain area involved in forming memories, is most susceptible to stress hormones. Even a single severe stress episode can destroy newly created neurons in the hippocampus. Chronic stress can kill off old neurons in the hippocampus. Research on post-traumatic stress after 9/11 terrorist attacks shows that hippocampus of some survivors had shrunk to the size of those elderly people with dementia.

Robert Sapolsky, a neurologist at Stanford University and author of *Why Zebras Don't Get Ulcers: An Updated Guide to Stress, Stress-Related Disease and Coping*, observes that, while a zebra will turn off the stress hormones after escaping from a lion, modern humans not only produce too much cortisol in response to everyday alarms but cannot turn them off afterwards. As we are not zebras, it's impossible to control stress. However, we can manage it easily. Try:

- Don't fight: Sapolsky's studies on baboons show that walking away from provocations is the personality trait linked reliably with lower levels of stress.
- Follow the old saying 'It's better to give than receive': Many studies have shown that there is joy in giving and it may have stress-reducing effect. In one study participants' brains were scanned when the received cash rewards in a computer game and when they donated their winnings to charities. The scans showed more activity in the award-related regions of the brain when the participants gave money than when they just got money.
- Laughter: Laugh a lot and take nothing seriously, a wise philosopher has said.
- Get enough sleep.
- Make friends to create a buffer against stress.
- Meditate.

84

Could we persuade your friends to buy this book?

If we take the liberty of dreaming that Robert Cialdini of Arizona State University, an internationally respected expert in the fields of persuasion and influence, was spruiking this book, he would most likely succeed in persuading your friends that it's worth buying. After extensive research on persuasion, he has distilled six rules of persuasion. In brief:

1. *Reciprocation*: Social norms obligate us to repay in kind what we have received. Freebies (in-store wine tastings or free doughnuts with coffee) from businesses expose us to products or services and create a feeling of 'indebtedness' leading us to buy something.
2. *Consistency*: Using the force of another potent human motivation: the desire to be, and to appear, consistent.
3. *Social validation*: We want to be part of the scene: one fundamental way that we decide what to do in a situation is look what others are doing or have done there.
4. *Liking*: People prefer to say yes to those they like. We are also swayed by people who are similar to us in appearance, hobbies or behaviour.
5. *Authority*: We fall for authoritatively sounding claims such 'scientifically proven', 'tested in university laboratories' and 'recommended by dentists'.
6. *Scarcity*: Numerous studies show that items and opportunities become more desirable to use as they become scarce. Scarcity can be manufactured by

salespersons: 'They are flying out of the store. This is the last one we have left in stock.'

In persuasion, does a person's sex make a difference? There is no clear-cut answer. However, on the matter of online persuasion, research by Rosanna E. Guadagno of the University of Alabama shows that when a woman is trying to influence another she doesn't know, a face-to-face conversation works better than email because typically women get to know one another quickly in person. On the other hand, she advises, men are better off using email to focus on the text, not the persuader.

The Mysterious Mind

Hypnosis is no longer a stage act; it is now being used as a tool for studying the inner workings of the brain and for treating patients

85

You're getting sleepy ... very sleepy ...

Hypnosis started as a 'miracle cure' in 1775 when Franz Mesmer, a German physician, proposed that an invisible magnetic field flows through our bodies. If the flow were restricted somehow, it would cause physical and mental illnesses. He theorised that by passing magnets over the body, the fluid would be unblocked and the patient cured. He managed to cure some patients, all young women. In 1778 he moved to Paris where he became quite the rage. He would ask his patients to sit with their feet in a tub filled with 'magnetised water' while holding iron rods attached to the tub. Ethereal music played on a glass harmonica to induce hypnotic trance. He would than emerge from behind heavy drapes dressed in purple silk and holding an iron rod.

This showmanship infuriated Parisians so much that in 1784 King Louis XVI instituted a scientific enquiry into his bogus therapy, which concluded that the observed effects could be attributed to the power of suggestion (a kind of placebo effect). Ever since hypnosis (the name was introduced in 1842, until then it was known as mesmerism) has struggled for scientific respectability. However, it's now gaining some credibility as a tool for research and treatment.

Hypnotherapists do not use a swinging pocket watch or a pendulum; they give verbal guidance to the person to concentrate on particular images or ideas. When hypnotised, you are neither asleep nor unconscious. It's like when you are about to fall asleep. You are still aware of everything and do not lose control over your actions; but your subconscious mind is open to suggestion. The procedure is no more dangerous than listening to a lecture.

Some people are easily hypnotised than others. Ten to 15 per cent of adults can be easily hypnotised, while around 10 per cent are almost impossible to hypnotise. The rest are in between. If you can be easily hypnotised, you are likely to have imbalance in the efficiency of the brain's left and right hemispheres. During successful hypnosis brain's right hemisphere dominates temporarily. It is believed this state is much easier to bring about in people who have an imbalance on the efficiency of their two hemispheres, even when awake. Studies also suggest that during hypnosis there is less connectivity between the left and right hemispheres of the brain

Stroop effect (*page 42*) shows how the brain deals with conflicting information. Brain scans of people who took Stroop test (naming aloud the colours of words printed in incompatible ink colour; for example, word 'blue' printed in red ink) suggest that hypnotisable people show no activity in the visual area which usually decodes written words. And the anterior cingulated cortex, which sends a distress signal when things are going wrong, showed reduced activity. The results show how hypnosis can override the 'automatic' processes in the brain for deciding what to do in the face of conflict. Critics of hypnosis say that these changes are not unique to hypnosis; similar changes have been observed among awake people.

Medical studies on hypnosis are small, but they show that it may help in management of chronic pain and reducing stress and anxiety. It also seems to work very well in the treatment of irritable bowel syndrome. Critics of hypnosis say the positive effects of hypnosis probably result from people's expectation about hypnosis, not from the hypnotic state — the placebo effect.

The exact mechanism behind hypnosis has yet to be fully understood, but it has come a long way from the days of purple-robed Dr Franz Mesmer who got his doctorate for a thesis on how the gravity of various planets affects health.

86

Why we love dark side of things

Some people ignore horror movies likes *Scream*, *The Shining*, *Halloween*, *Nightmare on Elm Street*, *The Exorcist* and *Poltergeist*, but there are others who savour them. Why do some people enjoy intense experiences that are quite scary to others?

The right person to answer this question is American psychologist Frank Farley, much of his research has focused on what he calls type T personalities. 'There's a long history of people being intensely curious about the "dark side", and trying to make sense of it,' he says. 'Through movies, we're able to see horror in front of our eyes, and some people are extremely fascinated by it. They're interested in the unusual and the bizarre because they don't understand it and it's so different from our everyday lives.' Most of the time, 'the unusual and the bizarre' in movies are paranormal or supernatural things.

Type T personalities are thrill-seekers, risk-takers and rule-breakers. Not all seek thrills in movies. Farley's model of T 'positive' personality can account for involvement in entrepreneurship, extreme sports or science in arts. He believes Einstein was a big-time risk-taker: 'When he was pounding his theories, he was beyond the crowd, way out there in the realm of uncertainty and creating whole new visions — that's T-positive. T types tend to be natural born rule-breakers. They are very innovative.' T 'negative' personalities, on the other hand, embrace crime, violence or terrorism 'for the thrill of it'.

Back to scary movies, they attract audiences by immersing them in nearly two hours of fear, disgust, terror and depravity. How can the general assumption that humans pursue pleasure and avoid pain be reconciled with people's decision to engage in experiences known to elicit negative feeling? Psychologists offer two explanations: (a) viewers are not actually afraid, but excited by the movie; and (b) they are willing to endure scary moments to enjoy a euphoric sense of relief at the end. In their paper, 'On the consumption of negative feelings', American researchers Eduardo B. Andrade and Joel B. Cohen say that people can experience both fear and euphoria at the same time. In other words, they are happy to be unhappy. 'Within a certain range, the most pleasant moments of a particular event may also be the most powerful,' they say, comparing those moments to the thrill and fear of extreme sports

One explanation for differences between type T and non-T personalities is that the brains of thrill-seekers, risk-takers and rule-breakers need more dopamine, the neurotransmitter that provides a sense of reward and pleasure, while the brains of mild-mannered, risk-averse and rule-followers are already soaked in dopamine and they don't have to do a thing to seek a sense of reward and pleasure.

Can someone live with only half
a brain; or with a hole in the head?

87
Strange but true

In the 1880s Friedrich Goltz, a prominent German physiologist, performed a dramatic operation — hemispherectomy — in which he removed half of the brain of a dog. The dog retained reasonably good ability to see and walk and showed no apparent change in intelligence and personality. 'I succeeded in observing for fifteen months an animal in which I had taken away the whole left hemisphere,' he claimed in a research paper in 1888. 'A dog without a left hemisphere can still move all parts of his body.' He wanted to disprove the prevalent view that the brain was made up of organs with specialised functions.

Walter Dandy, an American surgeon, pioneered hemispherectomy on humans when in 1923 he performed the procedure on a patient with brain tumour in the right hemisphere. The man lived comfortably for more than three years after the operation. In 1938 Kenneth McKenzie, a Canadian neurosurgeon, was the first to perform hemispherectomy in a case of debilitating seizures that could not be controlled by medication. The operation was success, and today hemispherectomy is a well-established procedure for the treatment of certain kinds of epilepsies in children.

In 2003 an American study confirmed the lasting benefits of hemispherectomy. The study showed that the quality of life of 111 children with chronic, severe seizures greatly improved after the procedure. There was a downside: the children had partial paralysis on the side opposite the removed portion. However, most adapted to their handicapped side so well that they could walk, run and play the piano, golf, ping-pong — some could even dance and skip.

'When half of the brain is bad, it's better to take it out,' asserts Eileen Vining, an American neurologist and a co-author of the 2003 study, in the prestigious medical journal *Lancet*. 'This allows the remaining hemisphere to function more normally and often regain function that was lost in the constant seizures.' Children's brains are plastic — they have the ability to change — and the remaining portions of the brain overtake most of the functions of the missing side.

Yes, half a brain is better than a whole brain.

The idea of deliberately drilling a hole roughly the size of a man's watch into the skull to heighten consciousness seems barbaric, but archaeologists say that trepanation (from a Greek word meaning 'to bore') is the oldest surgical practice. A trepanned skull found in France was estimated to be 7000 years old. In the fifth century BC, the Greek physician Hippocrates, regarded as the father of Western medicine, wrote detailed instructions on how to perform trepanation to relieve pressure on the brain caused by disease or trauma. In some cultures it was practised to release evil spirits. This mix of magic and medicine has always fascinated people and the practice has not died out completely.

The most famous modern case of trepanation was recorded — literally on a film — in 1970. Amanda Fielding, a 27-year-old British painter, filmed herself performing self-trepanation. The film shows her standing in front of a mirror wearing a white frock. She takes a dentist's electric drill and starts drilling through the front of her shaved forehead. Fielding (now Lady Neidpath, she runs the Beckley Foundation, which does research on consciousness and its changing states) says that having a hole in her head allows more blood to reach her brain, which increases brain metabolism and expands her consciousness.

Trepanation is a dangerous operation — there are risks of blood clots, brain injuries and infections leading to meningitis or death—and no self-respecting surgeon would ever dream of performing it. There are no known psychological benefits associated with this crude ancient procedure.

No, 'hole in the head' is not a third eye to total perception. It's a fallacy.

Are our brains hardwired
to know right from wrong?

88

Mind to morality

Consider the moral dilemma posed by this longstanding philosophical puzzle known as the trolley problem: A runaway trolley is rolling down the track towards five workmen who will die unless you pull a lever to divert it onto a different track, but there is another workman on that track. If you pull the lever, he will be killed. Is it morally right to pull the lever to prevent five deaths at the cost of one?

Suppose you are with this workman on a footbridge spanning the tracks. The only way to save five workmen on the track is to push him off the footbridge and onto the path of the trolley — supposing he is large enough to stop the trolley; you're not? Is it morally right to push the workman off the footbridge?

Harvard University psychologist Joshua Greene studies moral judgment and decision-making using behavioural experiments and brain scans when people ponder problems like the trolley problem. In the lever dilemma, most people say 'Yes', while in the footbridge dilemma the answer is 'No'.

Greene poses two questions: (1) What makes it right to kill one person to save five others in the lever dilemma, but not in the footbridge dilemma? (2) How does everyone know that it's right to pull the lever, but not right to push the workman off the footbridge?

The eighteenth-century Scottish philosopher David Hume said that reason alone is merely the 'slave of the passions'. We like to think our views on right and wrong are rational, he said, but they are based on sentiments rather than reason. Greene has developed a dual-process theory that attempts to understand the trolley dilemma; it also shows that Hume was onto something that science is unravelling now.

The dual-process theory says that both emotions (intuitive response) and reason (more controlled response) play crucial, and in some cases, mutually competitive roles. Differing responses to the trolley dilemma show the operations of at least two distinct psychological/neural systems. The 'Yes' response in the lever dilemma fits well with the utilitarian perspective: it's better to save as many lives as possible. This system appears to depend on dorsolateral prefrontal cortex, a region in the right side of the brain which becomes more active when people make utilitarian choices. The 'No' response to the footbridge dilemma is consistent with deontological or duty-based perspective which emphasises individual rights over utilitarian considerations. This response excites parts of the brain, medial prefrontal cortex and amygdale in the left side of the brain, which are linked to emotions and social thinking. This is Greene's theory. The debate about how we juggle reason and emotions to make moral decision is still continuing.

Ventromedial prefrontal cortex (VMPFC), a region of the prefrontal cortex located above our eye sockets, is also believed to be a central part of a network of brain regions involved in moral decision-making. This was one of the regions damaged in the famous case of Phineas Gage (*page 84*) whose personality changed from kindly to aggressive. Studies have shown that people with VMPFC damage are more likely than others to make utilitarian choices in moral dilemma — they are more likely to say 'Yes' in the case of the lever dilemma. The reason perhaps is that VMPFC plays an important role in sentiments such as guilt, compassion and empathy — the absence of which may make it easier to make utilitarian judgments.

Patricia Churchland, an American philosopher of neuroscience and author of *Braintrust: What Neuroscience Tells Us about Morality*, argues that morality originates in the biology of the brain. In her biological story the hormone oxytocin plays an important role. Oxytocin is released from the posterior lobe of the pituitary gland, a pea-sized gland located at the base of the brain. It also works as a neurotransmitter in the limbic system, the brain's emotional centre. An interesting insight into oxytocin came in the 1970s when studies showed that it played an important role in fostering bonding and monogamous behaviour in prairie voles (there are two types of these fury little rodents — prairie voles bond for life, montane voles are promiscuous). Subsequent studies have shown that blocking the receptors for oxytocin in prairie voles changes their behaviour completely — they no longer bonded with their mates and stopped caring for their young. This observation inspired Churchland to conclude that oxytocin was the essence of empathy allowing us to develop the trust in one another for the development of morality. Perhaps, she says, Hume might accept oxytocin as the germ of 'moral sentiment'. Her biological picture of morality also includes other elements, especially prefrontal cortexes.

The moral of these researches seems to be that moral rules are not unchangeable or eternal, like the speed of light, as we would like to believe. They are innate — born in our minds. How they are born is still open to debate.

There may be nothing unique about religious belief in the brain, but belief certainly calms the worried mind

89

Searching for God in the brain

Worship of some variety of supernatural power is common to all cultures, which suggests the human brain is hardwired for God and it comes preloaded with software for religious experiences. Not everyone agrees with this view, and neuroscience has no definitive answers. However, a number of studies show that the brain does seem to be predisposed towards spiritual and religious belief. This idea has led neuroscientists in search for the area in the brain — God spot — that control religious belief.

One contender for God spot is the temporal lobes, the part of the brain that lies around ears. It's where epileptic activity takes place. Epileptics suffering seizures in the temporal lobes report of having religious and spiritual experiences. Michael Persinger, a Canadian psychologist, has used the relationship between religious experiences and temporal lobe epilepsy to simulate such experiences in normal healthy people. For his experiments he has designed a God helmet. This modified motor cycle helmet produces a very weak rotating magnetic field over the temporal lobes when placed over on the subject's head. The subjects who are placed in a quiet chamber while blindfolded report being in the presence of a spirit (if they are strongly religious they are likely to interpret this presence as God). The famous evolutionary biologist and author of *The God Delusion*, Richard Dawkins, who is an avid atheist, has also tried the helmet without experiencing anything. Nevertheless, he thinks there could be an evolutionary advantage, not believing in God, but to having a brain with the capacity to believe in God.

After analysing the brains of volunteers — some religious, some, nonreligious — a team of scientists at the US National Institutes of Health has come to the conclusion that there is no God spot in the brain. The researchers used fMRI to study the brains of volunteers who had been asked to think about religious and moral problems and questions. They found that people of different beliefs all tend to use the same regions of the brain — the regions that are activated when the brain is involved in empathy, and in deciphering what other people might be thinking. 'The results show that, to the brain, religious belief is a lot like political belief,' says Jordan Grafman, the lead researcher. There is nothing unique about religious belief in the brain.

There is nothing unique about praying to God either; it's just another kind of friendly conversion. This 'friendly' image of God comes from neuroscientist Uffe Schjoedt of University of Aarhus in Denmark and his colleagues. These researchers scanned the brains of twenty devout Christians who first silently recited the Lord's Prayer and then a nursery rhyme. The same brain areas, typically associated with rehearsal and repetition, were activated. 'It's like talking to another human,' says Schjoedt. 'We found no evidence of anything mystical.'

If you've believe in God and worried about what you have read so far, relax. You will find relaxing much easier than non-believers. Thinking about God makes you calm under fire. At least, that's the theory of psychologists Michael Inzlicht and Alexa Tullett, both at the University of Toronto, who used EEG (electroencephalograph) to measure brainwaves of participants performing a Stroop task (*page 42*), which shows how your brain deals with conflicting information. The task was deliberately chosen to produce a lot of mental errors. The brain has a kind of built-in monitor, a region of the brain called anterior cingulate cortex (ACC), which sends a distress signal when things are going wrong. When the participants with belief in God were primed to think about God, either consciously or unconsciously, brain activity decreased in ACC. Interestingly, when non-believers were unconsciously primed with God-related ideas, their ACC activity increased. 'Religion seems to act as a palliative for believers,' says Inzlicht. 'It buffers them against the pains of everyday living.' Don't despair if you're a non-believer. Similar effect would take place with atheists with a meaningful belief system that provides structure and helps them understand their world. 'Maybe atheists would do better if they were primed to think about their own belief,' remarks Inzlicht.

It's all about the power of belief. Belief is one of the most powerful medicines. Physicians have known for long that sugar pills disguised as medicines — placebos — can help some patients. Thalamus, a region of the brain that acts as the gatekeeper by relaying sensory information, releases pain-reducing chemicals such dopamine after placebo is given. Believing in a cure not only makes you feel better, it can lead to dramatic bodily changes.

Consciousness is the most obvious but acutely private matter of our minds; it cannot be conveyed to someone else. Does it emerge from the brain or does it require an immaterial soul?

90

The ghost in the brain

Until recently scientists ignored the study of consciousness, our personal awareness of the world and ourselves. The problem was considered either philosophical or too difficult to study experimentally. But most now believe that consciousness is likely to be explainable as the outcome of the interactions of the brain's neurons.

Sometimes described simply as awareness of awareness, consciousness is the essence of what is to be human. Most of what the brain does, it does outside our conscious awareness. We are largely unconscious of processes that create our perceptions, memory, thoughts, feelings, motivations and behaviours.

Francis Crick, who shared a Nobel Prize with James Watson for the discovery of DNA's structure in 1953, should receive much of the credit for the current scientific interest in looking for an explanation of how processes in the brain creates consciousness awareness. In 1990 he and Christof Koch, a young neuroscientist who collaborated closely with Crick, rejected the belief of many of their colleagues that consciousness cannot be defined, let alone studied. Consciousness is a legitimate subject for science, they declared. Crick died in 2004 but Koch is still actively involved in research on consciousness.

According Crick and Koch, one cannot hope to achieve true understanding of consciousness by treating the brain as a black box (an object whose internal structure is irrelevant). Only by examining neurons and the interactions between them could scientists accumulate the kind of knowledge required to create a scientific model of consciousness. Crick and Koch believe that, though there are many possible approaches to the problem of consciousness, they have focused on 'visual awareness rather than other forms of consciousness, such as pain or self-awareness, because humans are very visual animals and our visual awareness is especially vivid and rich in information.'

David J. Chalmers, an Australian philosopher, believes that philosophy must bridge the 'explanatory gap' between a physical theory of consciousness and our subjective experience. 'I got into this field to try and understand the problem of how a physical system like a brain could also be a conscious being with subjective experience,' he says.

Chalmers makes a distinction between the 'easy' problems and the 'hard' problem of consciousness. His list of easy problems includes: How is it that a brain can discriminate information from the world? How is it that it can bring it together in the brain and integrate it? How is it that the brain or a human being can verbally report their mental states? How is it that we bring information to bear in controlling our action?

Chalmers does not consider the easy problems as trivial problems, but he believes that continued work in cognitive psychology and neuroscience will answer them. But the hard problem would still remain: how physical processes in the brain give rise to subjective experience. 'Why is that physical processing in the brain, no matter how sophisticated, should give rise to any subjective inner life at all, why couldn't that have all gone in the dark?' he says, 'That's the real mystery.'

To illustrate the distinction between the easy problems and the hard problem Chalmers uses a thought experiment devised by the Australian philosopher Frank Jackson: Mary, a neuroscientist in the 23rd century who knows everything there is to know about how the brain processes colour, has lived all her whole life in a black-and-white room. Mary does not know what it is like to see a colour such as red. 'It follows that there are facts about conscious experience that cannot be deduced from physical facts about the functioning of the brain,' says Chalmers.

Daniel C. Dennett, an American philosopher, is the leading critic of the hard problem. His 'multiple drafts' theory says that consciousness is not a unitary process but rather a disturbed one. The brain is a kind of hypothesis-making machine, constantly throwing up new 'drafts' of what is going on in the world. Sequential timing of events breaks down at extremely small time scales within the brain, and the events that make up consciousness cannot be ordered. There is no central place in the brain where everything is presented or decisions are made. 'Mental states do not become conscious by entering some special chamber in the brain,' he stresses.

The rigid boundary between the easy problems and the hard problem is fast disappearing as neuroscientists pile up evidence that every aspect of the consciousness can be traced to the brain. 'And when the physical activity of the brain ceases, as far as anyone can tell the person's consciousness goes out of existence,' says Steven Pinker, author of the bestselling book, *The Language Instinct*. He believes that whatever the solutions to the easy and hard problems turn out to be, few scientists doubt that they will locate consciousness in the activity of the brain. But how the processes of the brain translate to consciousness is still a mystery to scientists. Even if they unravel this mystery, will it really explain consciousness? There are many who still hold to the view that our minds are more than brain and therefore consciousness will remain the ultimate mystery. They are uncomfortable with the idea of reducing their minds to pieces of tissues and their private ideas and feelings to pixels on computer screens.

Nevertheless, Koch is confident that 'pixels on computer screens' will eventually help us to solve the riddle of consciousness as new theories of consciousness, based on information science and mathematics, can describe what characteristics a network of neurons would have to be considered conscious.

Ray Tallis, a prominent UK neuroscientist, disagrees. He believes that any explanation of consciousness will always remain incomplete — or unreliable — because science can only do its work by discarding the content of consciousness. Science begins when we escape our subjective, first-person experiences into objective measurement. Thus measurement takes us further from subjective consciousness to a realm where things are described in abstract but quantitative terms. Our failure to explain consciousness in terms of neural activity inside the brain is not due to technical limitations which can be overcome, he argues, but due to the self- contradictory nature of the task.

Science has yet to capture consciousness.

Déjà vu is the eerie experience of having been somewhere or done something before even though there is evidence to the contrary

91

The feeling of familiarity

You walk into a new place or find yourself talking to a stranger, and suddenly you have this uncanny sense of familiarity, yet you know you they are like nothing on earth.

Surveys show that almost two-thirds of individuals have experienced at least one déjà vu ('already seen' in French) in their lifetime, and these individuals typically had multiple déjà vu experiences. The relatively brief experiences — they last only 10 to 30 seconds — are usually triggered by a physical situation, although spoken words alone sometimes can cause the illusion. 'Personal reactions to déjà vu are more positive than negative, and people typically indicate that they are surprised, curious, or confused when they experience the illusion,' remarks Alan S, Brown, a psychologist at Southern Methodist University in Texas.

Déjà vu experiences are memory illusions caused by false recollection, but none of the thirty or so scientifically plausible explanations provided by psychologists since the 1880s has fully explained the mechanism of déjà vu. A groundbreaking experiment by a team of neuroscientists at the Massachusetts Institute of Technology's Picower Institute for Learning and Memory has now suggested that the familiar feeling comes from a conflict between two parts of the brain.

For their study, the MIT team used genetically engineered mice that lacked a component of memory called pattern separation. It's the brain ability to separate very similar and closely related experience. 'You can have dinner at a restaurant with the same group of friends on two nights, and remember them as two distinct experience based on small differences — the topic of conversation, the table you sat at, the entrée you chose,' explains Thomas McHugh, a team member. Several years ago, Susumu Tonegawa, who won the 1987 Nobel Prize for Physiology or Medicine for his work on the genetics of immunity, had uncovered a related component of memory called pattern completion. This mechanism enables you to recall complete memories based on just a single clue ('When we were at the restaurant last week'). Tonegawa's discovery helped identify the gene that regulates pattern separation.

Normal and genetically altered mice were first placed in a box which gave them a mild foot shock. They all reacted by freezing. Then, they were placed in a very similar box with no shock. Within weeks, the normal mice quickly figured out the difference between the two boxes. Genetically altered mice, on the other hand, became confused and froze in fear in both boxes. These mice were unable to discriminate between different spaces as they lacked pattern separation involved in identifying new and old space. Researchers believe that 'place neurons' fire to provide a sort of blueprint for a new space we encounter. The next time we see the space, those same neurons fire. Thus we know when we have been somewhere before and do not have to relearn our way around.

'Déjà vu in humans could at least be partially explained by a failure of the pattern separation circuit — the misattribution of a new experience to a previously stored experience,' says McHugh. 'The eeriness comes from the fact that another part of the brain is telling you: This feeling of familiarity? It's actually not true at all.'

92

Perception beyond belief

Bertrand Russell, the English philosopher and mathematician who won the 1950 Nobel Prize for Literature, once said, 'Man is a credulous animal, and must believe something; in the absence of good grounds for belief, he will be satisfied with bad ones.'

He would have certainly categorised paranormal beliefs as 'bad ones'. The paranormal is beyond normal; the word comes from the Greek prefix *para* meaning 'beyond'. Paranormal powers are normally divided into two types: extrasensory perception (ESP) and psychokinesis. Parapsychologists who study these alleged phenomena use the word 'psi' (from the Greek *psyche*, mind) to refer to both ESP and psychokinesis; and psychics are people who are said to possess powers of psi.

How can one become a psychic? Some people believe they were born with psychic powers. Some claim that they gained these powers after an accident or a traumatic event. Others try to learn them. Those who are not psychic experience the world through five senses: vision, sound, touch, taste and smell. Our normal perception is limited by our senses. For example, our sense of sound is limited to a frequency range between 20 and 20,000 hertz (cycles per second), and our vision is limited to wavelengths from about 400 to 700 nanometres.

ESP has no such limitations, as it is supposed to be gained without the use of the usual five senses. This sixth sense can be the ability: (1) of two minds to communicate through an unknown channel (telepathy); (2) to be aware of an unknown object or event (clairvoyance); (3) to know future events (precognition); or (4) to know past events (retrocognition).

After decades of research parapsychologists cannot even explain how ESP works. As many snake-oil theorists look for evidence in quantum physics these days, it's not surprising that some parapsychologists try to relate ESP to quantum nonlocality. Imagine this book is sitting on a table in front of you. To move it you have to touch it. We can only affect objects we can touch. We experience the world as local. Newton's discovery of gravity introduced the idea of action at a distance or what we call classical nonlocality. In the quantum world nonlocality or action at a distance comes from quantum entanglement, in which the information between the two particles is somehow transferred.

Quantum entanglement is one of the bizarre features of quantum physics. All elementary particles such as electrons and photons vibrate. Consider the case of two electrons. Placed together they vibrate in unison. Place them apart, as far as another galaxy, and if you vibrate one of them, the other will immediately know the nature of its partner's vibrations and dance to the same tune. Somehow the information between the two electrons is being transferred. Einstein called it 'spooky action at a distance' because the transfer of information could only be explained by assuming that it was travelling faster than light.

Could quantum entanglement be used to transfer information faster than light? The laws of quantum physics prohibit it, but it could be used to transfer information from one particle to another particle at a speed slower than that of light. In quantum teleportation, only the quantum state is teleported, not exact particles. So, quantum teleportation is not really instantaneously sending one particle, say an electron, from one place A to another place B. However, the quantum states of the electron at A and B are indistinguishable.

Parapsychologists say that our minds are physical objects and can therefore be described by quantum theory. In other words, like two quantum particles, information can be transferred between minds. So, telepathy can be explained as a quantum connection. Sounds convincing, but there is a major flaw in this argument: our brains are too chaotic to sustain a fragile state known as quantum coherence, which denies quantum entanglement.

Is ESP fact or fantasy? Science can neither prove nor disprove ESP. Some hold the view that if ESP could be supported by empirical evidence, it would no longer be paranormal phenomena. Is there a psychic out there who has won lottery after lottery? No one has ever succeeded in peeking a little way into the future. Until it can be demonstrated convincingly by objective researchers, ESP remains beyond belief.

Two Harvard University objective researchers, Samuel T. Moulton and Stephen M. Kosslyn, recently set out to resolve the paranormal phenomena debate. If ESP or such things exist, they occur in the brain. Moulton and Kosslyn scanned the brains of nineteen pairs of individuals (couples, romantic friends, twins) to assess whether individuals can have knowledge that does not come from normal perceptions. The participants viewed 240 pairs of photographs while inside an fMRI scanner. Each picture pair was randomly assigned a stimulus category — ESP or non-ESP. ESP stimuli pictures were also presented to the subject via three different forms of ESP: telepathically (shown simultaneously to the subject's partner in a separate location); clairvoyance (displayed on a computer located outside the subject's field of vision; and precognition (shown to the subject at a later time).

The researchers found that the participants responded identically to both types of stimuli. The result supports the null hypothesis (the opposite of the hypothesis being tested). 'We didn't find anything, but we didn't find anything in an interesting way,' the researchers say.' They agree that while null results can never be used to conclusively disprove that ESP doesn't exist, but they are happy to have done their bit in settling an age-old debate that belongs on the fringes of science.

93
Mind over matter

Telekinesis (also called psychokinesis) is the alleged ability to move objects by mental effort alone. Simply put, it is mind over matter. Acknowledged in many cultures since the dawn of history, telekinesis gained enormous publicity in the early 1970s when the Israeli psychic Uri Geller appeared on television around the world claiming to bend spoons and other metal objects, apparently with the force of his thoughts. The Force wielded by the Jedi Knights in the popular *Star Wars* movies is also telekinesis.

Geller's spoon-bending trick has been shown as a quick sleight of hand, not the result of psychic powers. The Force is simply science fiction. But no amount of demonstrable fraud or fiction can dissuade the true believers in the power of paranormal.

Telekinesis is inconsistent with the laws of physics, which have been proven beyond any reasonable doubt, creating a problem for scientists who want to study it. Another problem, says American theoretical physicist Michio Kaku, is that scientists are easily fooled by those claiming to have psychic power. 'Scientists are trained to believe what they see in the lab,' he says. 'Magicians claiming psychic powers, however, are trained to deceive others by fooling their visual senses. As a result, scientists have been poor observers of psychic phenomena.' Einstein put it in a subtle way: 'Nature hides her secrets through intrinsic grandeur but not through deception.'

These views didn't deter Dr Robert G. Jahn, a professor of engineering, from founding the Princeton Engineering Anomalies Research laboratory at Princeton University in 1979. Until it was closed in 2007, the laboratory carried out research in 'micro telekinesis' to prove that thoughts can alter the course of events.

The event most studied by Dr Jahn's laboratory was the flipping of coins. The theory of probability tells us that the when we flip a coin there is a 50 per cent chance of getting heads or tails. Dr Jahn's team designed an experiment which was an equivalent of a coin flipper. Volunteers would sit in front of an electronic box that flashed random numbers just above or just below 100, but had no physical connection to the machine. They would then try to influence the outcome by thinking 'high' or 'low' to produce a higher or a lower number than it should be by chance.

When the researchers looked for difference between the machine's output and random chance after more than two million 'coin flips' over two decades, they found the effects very small but amazing. The volunteers 'thoughts' were roughly altering one number in 1,000. In other words, if you had a coin flip, psychokinesis could affect one of those coins flips if you flipped a coin thousand times.

Scientists dismiss this and other similar data on the grounds that such minute differences could have been caused by subtle, hidden biases in the experiment's design. If you disagree, you might try telekinesis in front of a poker (or slot) machine. Think positive and be nice to the machine and whisper in a soothing voice, 'Sure, you can do it, sweetheart', and it will come up with the jackpot.

Of course, telekinesis enthusiasts insist that it's possible, in varying degrees, for the human mind to influence the physical environment. However, No one has ever produced a satisfactory theory to explain their claim.

We take it for granted that 'self' conforms to the borders of the physical body, but ...

94

Can mind leave the body?

Out-of-body experiences — when a person sees his or her body from a location outside the physical body — are fairly common and researchers estimate that about 15 to 20 per cent of the population have had at least one out-of-body experience during their lives. People usually have such experiences when they are awake, not during sleep. The experiences can be spontaneous, but most often they occur during periods of high stress, serious illness, acute trauma, abuse of hallucinogenic drugs, hypnosis, meditation or religious prayers. There is no evidence that they are caused by mental illness.

Most out-of-body experiences have many common features. Persons having the experience observe their physical body lying motionless and are able to move around it. They perceive that they have another body, a double of the physical body; this ghostly or transparent body can pass through walls and other solid objects. Sometimes they see that their 'astral' body is connected to their physical body by a silvery 'umbilical' cord. As in near-death experiences, some experiences begin with the entry into a dark tunnel which has a bright, white light at the end of it. Most experiences are brief and end with a sense of pleasure and liberation.

Scientists do not doubt that people do have out-of-body experiences, but they do not agree that experiences are psychic, paranormal or mystical. Some neuroscientists have tried to induce elements of illusory out-of-body experiences in healthy volunteers by using electrodes to turn off the temporoparietal region of the brain (a region of the brain where temporal and parietal lobes meet; people with damage to this region are no longer able to map body's position in space and, for example, have difficulty negotiating their way around a house). These laboratory experiments have shown that perceptual illusion can be induced in which volunteers experience that their centre of awareness, or 'self', is located outside their bodies and they look at their bodies from the perspective of another person. These and other experiments suggest that consciousness of a self in our body is based on processing of various sensory inputs in the brain, and the brain's representation of the physical body is changeable and can be modified by information from the senses.

The idea that soul or spirit can leave the body is an ancient one and is found in many cultures. For many, the most appealing explanation for out-of-body experiences is that the mind or soul in fact leaves the body. This explanation raises the question of how nonphysical mind could observe and reflect when it has left the physical body. Our mind depends upon the functioning of the brain in our body.

Are out-of-body experiences a purely psychological phenomenon, involving no soul or self leaving the body; or, are they a combination of imagination and extrasensory perception? Science has yet to provide a definitive explanation for this weird phenomenon but all the evidence shows them to be the misfiring of neurons, not paranormal or supernatural phenomena.

The other-worldly event known as a 'near-death experience' is not a paranormal event. The answer lies in the brain

95

There is no light at the end of the tunnel

What happens when we die? Obviously nothing as death is the final frontier and we have simply ceased to exist.

People who have come very close to clinical death and have survived do not agree. They tell amazingly similar stories about their experiences. The first stage of their 'life after life' is the feeling of immense peacefulness and the absence of pain and fear. They then somehow leave their physical body and find themselves looking down upon it. They continue to rise above their body and enter into a dark tunnel. Their peaceful journey ends when they see a light at the end of the tunnel. The distant, golden light is welcoming and some regard it as a supernatural presence of some sort. They think they have reached the boundary between life and afterlife. Some even recall speaking to their dead relatives or an encounter with certain aspects of their lives. What follows is the realisation that they have to leave this afterlife and then they wake up.

The details of such experiences may also have a religious or a cultural aspect; for example, Christians tend to see Jesus in the light, while Hindus see Yamdoot, the messenger of death. But the core characteristics of the experiences are the same across all cultures.

People who have returned from the brink of death often see their near-death experiences as paranormal experiences. There is a soul or psyche dwelling in the physical body. When we die this immaterial essence leaves the body and travels to another world.

Naturally, scientists disagree. They look for explanations in the brain, not in the supernatural world. They say that the brain surges with activity just before death. The retrospective analysis of brain activity of critically ill patients as they were removed from life support has shown that there is a significant spike in brain activity at or near the time of death. This increase in brain activity, which lasts from 30 to 180 seconds, could explain near-death experiences.

Some researchers associate near-death experience with the temporal lobes, the part of the brain that lies around ears. Studies suggest that electrical activity in the right temporal lobe is involved in mystical and religious experiences. Investigations of temporal lobe activity in people who had near-death experiences during life-threatening events have revealed that such a people have more temporal lobe activity than normal people. Everyone agrees that near-death experiences have some basis in normal brain function gone off course.

Others seek answers in our three states of consciousness: awake, rapid-eye-movement (REM) sleep and non-rapid-eye-movement (non-REM) sleep. People who have near-death experiences are more likely to get stuck between REM state (the state in which dreaming occurs) and waking. 'Lucid dreams are among the closest things we know of to a near-death experience,' says Kevin Nelson, an American neurophysiologist. He suggests that near-death visions of tunnels have nothing to do with near-death experience. They are due to a lack of blood flowing to the eyes. When the near-death experience is about to end, the brain enters REM state and the visual system becomes active giving the impression of light at the end of the tunnel.

Scientists have yet to fully explain the cause of near-death experiences, but it does not mean that these experiences are supernatural. 'People like to say that these experiences are proof that consciousness can exist outside the brain, like a soul that lives after death,' says Nelson. 'I hope that is true, but it is a matter of faith; there is no evidence for that.'

And these experiences do not happen when people have been declared clinically dead. In fact, these people had a cardiac arrest: the heart stops pumping which restricts the flow of blood to the brain. The brain damage occurs only after about 30 minutes when the flow has been reduced by 90 per cent or more. Cardiac arrest is reversible if treated early. People who had a near-death experience were not really dead.

So-called alien-abduction stories may seem fantasies to us but they play a role in the psychological lives of those who believe in them

96

When your brain has been abducted by aliens

There is no hard evidence that there is intelligent life beyond Earth, but still some people believe that aliens have visited the planet, and some even believe that they were actually abducted by aliens. The abduction stories of these people have a remarkable degree of similarity and consistency: Your naked body is floating on a high-tech table inside a round bright saucer-like object and being subjected to a painful invasive medical examination by aliens with large heads, slanted wrap-around black eyes, either grey, white or green skin and no hair or nose...

People who believe in alien abduction have a tendency to fantasise and to hold to unusual beliefs and ideas. They also believe in things like ESP, astrology, tarot, channelling, auras and crystal therapy. Susan Clancy, author of *Abducted: How People Come to Believe They Were Kidnapped by Aliens*, says these people are not crazy, but they have in common a rash of disturbing experiences for which they are seeking an explanation. For them, alien abduction is the best fit. 'Many of us long for contact with the divine, and aliens are a way of coming to terms with the conflict between science and religion,' she says. 'I agree with Jung: extraterrestrials are technological angels.'

Some psychologists have tied the phenomenon to sleep paralysis, a condition where the usual separation between sleep and wakefulness gets out of synchronisation. Sleep occurs when the body is in the dream phase of sleep and it disconnects from the brain. The brain is either awake or semi-awake but the body cannot move. At that point the sleeper often 'sees' shadowy creatures, 'experiences' levitation and 'feels' painful sensations throughout the body, explains Kazuhiko Fukuda, a Japanese psychologist. Such experiences match with the accounts of people claiming to be victims of alien abduction.

There are other psychological explanations as well. Some psychologists believe that alien abductions and other mystical and psychic experiences may be linked to excessive bursts of electrical activity in the temporal lobes. These lobes — one on each side of the brain, located near the ears — control hearing, speech and memory. Some argue that alien abduction may be disguised memories of sexual abuse. Or, they may be false memories. People can and do make powerful memories and these memories can take on a life of their own. All abduction stories have been recalled under hypnosis. Hypnosis makes people susceptible to creating memories of things that were suggested to them or things they just imagined.

US psychologist David V. Forrest offers two medical hypotheses. His 'strong' hypothesis is that the abductees are recovering memories of actual surgery — the memories may be actual recall of the operating room before losing consciousness or they could be memories from childhood filtered through childhood amnesia. His 'weak' hypothesis is that abductees are confabulating media conceptions of aliens with images of surgical and medical procedures generally, images that they may or may not have experienced personally. (*More on confabulation on page 119.*)

Whatever may be the right explanation for accounts of alien abduction, why are their stories so similar? Anthropologists point out that individual illusionary experiences conform to cultural patterns. Alien abductions occur mainly in the United States where people are familiar with alien references through supermarket tabloids, books, movies and TV shows. Abduction accounts became appearing from 1962 when alien abduction also began appearing on TV and the movies. During the witchcraft craze in medieval Europe, for example, many people reported being carried away by witches on broomsticks and being seduced by demons. Today's counterparts talk of being picked up by flying saucers and being forced to perform various forms of unwilling sex by their bug-eyed kidnappers.

Touch wood (or knock on wood) before you read about...

97

Science of superstition

Rituals like not crossing the path of a black cat or not walking under a ladder or saying 'God bless you!' when someone sneezes (to stop their soul flying away) may seem like medieval mumbo jumbo, but they are still very much with us. What purpose superstitious beliefs serve, and where they come from?

'When you are blue, just knock on wood,' so goes a song from the classic 1942 movie Casablanca. Studies show that superstitious behaviour increases under stress. A possible explanation is that stress reduces our sense of control and to regain that control we engage in superstitions. To test this hypothesis, psychologist Giora Keinan of Tel Aviv University measured 'touch woods' (or 'knock on woods') made by participants during interviews which included high-stress and low-stress questions designed to elicit the ritual. He found that the frequency of the ritual increased under conditions of stress. He also found that there was a greater difference between the number of gestures between the high- and low-stress conditions participants who felt a greater desire for control.

As a sports fan psychologist Lysann Damisch of University of Koln knew about the rituals of some famous sportspersons; for example, golfer Tiger Woods wears a red shirt when competing on Sunday, usually the last day of the tournament; and former basket ball player Michael Jordan wore his college team shorts underneath his NBA uniform. Stress didn't seem the cause of their quirks. Damisch wondered why they were doing so. She thought a belief in superstition might help people do better by improving their self-confidence. She and her colleagues designed a number of experiments to activate good-luck-related superstitions among participants via a common saying, action or a lucky charm. The results showed that participants who had lucky charms (all kinds of items from old stuffed animals to wedding rings to lucky stones) did better than participants who didn't have lucky charms. Similarly, wishing participants good luck ('I press the thumbs for you,' the German equivalent of 'keep your fingers crossed') also improved participants' success. The irrational idea that a lucky charm may help improve your performance now seems rational.

Scientists dismiss superstitions as creations of irrational minds, but they are still interested in finding out where they come from. Darwin had no time for superstitious beliefs like spilling salt or breaking a mirror brings bad luck; however, evolutionary biologists Kevin Foster of Harvard University and Hanna Kokko of the University of Helsinki believe superstition may have evolved to help us survive. Superstition is where we believe that one thing has caused another, they say, even if there's no evidence for it. The tendency to falsely link cause to effect — a superstition — is occasionally beneficial. Early humans displayed behaviours that implied a causal relationship that wasn't there. For example, they might associate rustling grass with the approach of a predator. Even if the wind caused the sound, the researchers say, but if a pride of lions is coming there's a huge benefit to not being around. Foster and Kokko's assertions are based on a mathematical model of situations in which superstition is adaptive. 'Humans are heavily affected by culture as well as evolution,' Foster says. 'Nevertheless our analysis suggests that cultural effects are shaped by an evolved tendency to really associate events, so readily that individuals often make superstitious mistakes.'

Experiments by psychologists Jennifer Whitson of the University of Texas and Adam Galinsky show how the brain's lack of control could lead to superstition. In a series of experiments they manipulated participants' sense of control. In one of the experiments the participants were shown twenty-four 'snowy' photographs, half of which contained hidden images such as a horse, a chair or the planet Saturn, while the other half just contained grainy random dots. Although all participants saw hidden figures, the participants who had developed a sense of lacking control by working on an earlier experiment saw more figures in the photographs that had no embedded images. Our brains are programmed to creating meaning out of patterns that we think we see in nature. When the brain loses control it instinctively seeks out patterns to regain control — even if by imaging patterns. We believe in things that aren't there because our brains like filling blanks.

Neuroscientists have found a surprised suspect for superstition: dopamine, a chemical that controls the brain's sense of reward and pleasure. In experiments by neurologist Peter Brugger of University Hospital in Zürich, two groups of participants were asked to distinguish real faces from scrambled faces and real words from made-up words as the images were shown briefly on a screen. Believers saw more faces and words when there was none, than did sceptics. The participants were then given a drug to increase the levels of dopamine in the brain. Both groups made more mistakes under the influence of the drug, but the sceptics' threshold fell and they saw more scrambled photos and words as real. 'Dopamine seems to help people see patterns,' says Brugger. This suggests that superstition is associated with high levels of dopamine in the brain. The result is not yet conclusive.

'In all superstition,' Francis Bacon noted in his 1612 essay on superstition, 'wise men follow fools; and arguments are fitted to practice, in reverse order.' Follow wise scientists as they continue their research on superstition.

We are suckers for weird beliefs because the brain allows and even encourages them

98

Beliefs in bizarre

Alien abduction, ancient astronauts, astrology, crystal healing, ESP, magnetic therapy, quantum healing ... the list of absurd things is long, and the number of people who not only believe in such rubbish but vigorously defend them is very large. There is absolutely no evidence, scientific or otherwise, to support their beliefs. Ockham's razor — the simplest explanation is most likely to be right — is alien to believers in UFOs; science is crystal clear to practitioners of crystal healing.

Why does the human brain allow and even encourages beliefs that defy reason? The answer probably lies in your parietal lobe — a mass of tissues at the top of the brain which process sensory input and distinguishes where the body ends and the material world begins. During intense prayer or meditation, the parietal lobe powers down. Unable to find the dividing line between self and world you experience the sense of having lost your worldly moorings. You feel connected to the 'other world'. The parietal lobe's ability to go quiet may encourage other beliefs that bring sense of connection.

Brain imaging these days can reveal many mental acts. A trio of American neuroscientists, Sam Harris, Sameer Sheth and Mark Cohen, just did that to find out how the brain differentiates between belief and disbelief. Their simple and ingenious experiment involved presenting a series of written statements to participants while they were in the fMRI scanner. The statements were designed to be clearly true (belief), false (disbelief) or doubtful were from seven categories: mathematical, geographical, semantic, factual, autobiographical, ethical and religious; for example:

62 can be evenly divided by 9
Senegal borders Guinea
'Devious' means 'friendly'
Most people have 10 fingers and 10 toes
You have two sisters
It is bad to take pleasure at another's suffering
There is probably no actual Creator God

The results were fascinating. Response time to true statements was much shorter than responses times to both false and doubtful statements; however, there was no difference in response times between false and doubtful statements. Response times to acceptance of belief were similar, whether the participants made judgment in the highly emotional areas of ethics or religion or seemingly neutral area of mathematics.

Most strikingly, different brain regions lit up when responding to belief and disbelief statements. True statements activated areas of prefrontal cortex which are thought to play a role in decision-making, memory and fear, while false statements showed increase activity in an area called the anterior insula which helps to process fear, disgust and reactions to bad smells. This finding suggests that there are two distinct brain systems for belief and disbelief. The researchers suggest that belief or acceptance of a proposition as true has a pleasant and rewarding emotional tone. Disbelief or rejection of a proposition, on the other hand, is often associated with a feeling of discomfort and urge to avoid 'untruth'. 'When someone says something you disbelieve, it has a kind of emotional tone,' says Harris. 'Rejecting someone's statement as illogical or incompatible feels like something.'

Believing or doubting something controls our behaviour and emotions. While doubt tends to inhibit action, beliefs make it easier to arrive at a decision and act on it. Michael Shermer, publisher of *Skeptic* magazine, states that once we form beliefs, we maintain and reinforce them through five cognitive biases that distort our precepts to fit belief concepts:

Anchoring bias: Relying too heavily on one piece of information.
Authority bias: Valuing too much the opinions of an authority.
Belief bias: Accepting an argument on the believability of its conclusion.
Confirmation bias: Seeking supporting evidence, but ignoring discomforting evidence.
In-group bias: Placing more value on the beliefs of our peers.

Should you worry if someone you know believes in weird things? 'It can cause no harm,' says Stephen Law, author of *Believing Bullshit: How Not to Get Sucked into an Intellectual Black Hole*. 'But the dangers are obvious when people join extreme cults or use alternative medicines to treat serious disease.'

When the brain activity ceases,
we're dead. Is death to be feared?

99

Death is death

When our heart stops beating and there is no blood flowing to our brain, all brain activity ceases and we are declared clinically dead. Brain death is arguably the most accurate biological and philosophical representation of death. Heart failure in itself does not constitute death. Pumping of blood is important, but it only supports the functions of the brain. Simultaneous monitoring of heart rate and brainwaves shows that it takes brainwaves 11 to 20 seconds to go flat when the heart stops pumping.

Is death instantaneous? No, doctors say that we take from a few seconds to a few minutes to go through the dying process. 'The last breath is taken, death takes hold and life is over,' says Thomas Kirkwood, author of *Time of Our Lives: The Science of Human Aging*. 'At this moment, most of the body's cells are still active. Unaware of what has happened they just carry on ... In a short while, starved of oxygen, the cells will die.'

When the cells are dead the tissues start decomposing and the body eventually reverts back to dust and only skeleton and teeth are left. Depending on environmental conditions such as temperature and humidity, it may be as quickly as two weeks or as slowly as two years.

What follows death? Do we experience our own non-existence? Do we suffer from being dead? Is there life after death? Can we know anything about afterlife? Do the souls of dead continue to exist in an afterlife world? Science is not concerned about these questions. They belong to philosophers and theologians.

The fourth-century BC Greek philosopher Epicurus, who believed that sensory enjoyment was the most important thing in life, was the first to argue that we must not fear death because we won't be able to experience our own non-existence. His followers were called Epicureans and they lived in a garden. The story goes that an inscription on the gate to the garden said: 'Stranger, here you will do well to tarry; here our highest good is pleasure.' Yet the life of community was simple and frugal. Their food and drink was mainly bread and water. Epicureans continued to flourish for another five centuries and exerted a considerable influence over all ancient philosophy and science.

Epicurus said that death is the primary cause of anxiety among human beings. He argued that death involves neither pain nor pleasure. The only thing that we should fear is pain. Therefore, we should not fear death. 'Death, the most awful of evils, is nothing to us, seeing that, when we are dead, death is not come, and when death is come, we are not,' he said. In other words, death is nothing to us, since while we exist, our death is not, and when our death occurs, we do not exist.

On the matter of afterlife, Peter Atkins, presents a scientific answer in his excellent book, *On Being: A Scientist's Exploration of the Great Questions of Existence*: 'Quite frankly, it is hard to credit that the extraordinary property of complexly interacting organized matter we call consciousness can survive the decomposition of that matter. Science, especially through psychology, shines its brilliant light on the afterlife and instead of illuminating it causes it to shrivel and die, revealing its core: anxiety.'

It's not to say that all scientists reject the notion of afterlife. In his later life the practical-minded Thomas Alva Edison tried to tackle the concept how soul could live after death. The great inventor is rumoured to have built, but never built, a 'Telephone to the Dead'. We can talk to the dead any time we like, but the hard part is getting the dead to talk back.

100
When our minds will merge with machines

Humans are still evolving and, according to some experts, probably quite rapidly. Computers are also evolving and the rate at which they are becoming smarter is also increasing rapidly. So where are we and our computers heading?

Computers are becoming smarter and smarter because the number of transistors that can be packed on a chip is increasing fast. To make us more intelligent, could the evolutionary process also pack more neurons in our brain or boost the rate at which these neurons make connections? This could happen in three possible ways: by increasing the size of the brain, by packing more neurons in the existing space, or increasing the speed at which neurons process information. These ways all face thermodynamic hurdles (the laws of thermodynamics ensure that you cannot get something for nothing). Neurons are energy hungry — as mentioned earlier, our brain uses about one-fifth of our body's energy resources. Increase in brains size by adding more neurons would certainly increase its processing power. A bigger brain means longer axons, the 'wires' that join up neurons, which would increase their demand on the body's energy. Packing more neurons in the same space would slow down exchange of information as smaller axons or neurons tend to fire randomly. Increasing the speed of connectivity also requires more energy. For these reasons, some neuroscientists suggest, our brains are already as smart as a neuron-based brain can be.

Ray Kurzweil, an American inventor and futurist, has an interesting take on this problem. He says that in the not-too-distant future technological change will become so rapid and profound, humans will merge with technology and biological evolution will become obsolete. This 'cutting-edge of evolution on our planet' will create something with smarter-than-human intelligence. He calls this new era 'the Singularity'.

In his bestselling book, *The Singularity Is Near: When Humans Transcend Biology*, Kurzweil says, 'As we get to the 2030s, the non-biological portion of our intelligence will predominate. By the 2040s it will be billions of times more capable than the biological part.'

Kurzweil divides the present and the future history of evolution into six epochs:

Epoch One: Physics and Chemistry. In this epoch, which started soon after the Big Bang, information was in atomic structure, especially in complicated, information-rich, three-dimensional structures of carbon atoms. Evolution, to Kurzweil, is a process of creating patterns of increasing order. The exquisite and delicate balance of the physical laws of the universe is precisely what is needed for the codification and evolution of information, resulting in increasing levels of order and complexity.

Epoch Two: Biology and DNA. In the second epoch, starting several billion years ago, matter transformed into 'life' and thus physics into biology. DNA provided the precise digital mechanism for storing information and keeping it as a record of the evolutionary experiments of the epoch.

Epoch Three: Brains. The third epoch started when early animals developed the ability to recognise patterns, which still accounts for the vast majority of activities in our brains. Like earlier epochs, this epoch also provided a revolutionary or paradigm shift in the evolution of information: organisms now could detect information with their sensory organs and process and store that information in neural patterns in their brains.

Epoch Four: Technology. The pace of evolution of human technology is quickening. This quickening is exponential (when we plot growth against time on a graph, a J-shaped curve shows exponential growth; a straight line shows linear growth); for example, it took 68 years from light bulb to transistor, 30 years to first personal computers and then only 14 years to the World Wide Web. 'We won't experience 100 years of progress in the 21st century — it will be more like 20,000 years of progress at today's rate,' he says.

Epoch Five: The Merger of Human Technology and Human Intelligence. When the Singularity arrives, the human-machines will 'overcome age-old problems and vastly amplify human creativity'. Does this mean no nukes, no wars and the human-machine race lives forever in harmony and peace? Kurzweil is not so sure: 'But the Singularity will also amplify the ability to act on our destructive inclinations, so its full story has not yet been written.'

Epoch Six: The Universe Wakes Up. In the last and dramatic epoch in 'the evolution of patterns of information', our civilisation will infuse the rest of the universe with its creativity and intelligence (we suppose if it doesn't destroy itself in Epoch Five).

Kurzweil's scenario is built largely upon recent exponential growth in computing power. In 1965 Gordon Moore, one of the founders of chipmaker Intel, made his famous prediction: the number of transistors on a computer doubles every 18 months or so. This prediction, now known as Moore's law, still holds true. But this exponential growth depends on many variables: cost of energy, changes in environment, and political and economic stability. It is probable that the current exponential growth in the power of technology can slow down, stop or even recede.

Another major problem for Kurzweil's theory is the realisation that infusion of humans and machines can happen only when human-machines can circumvent the limit that the speed of light imposes on the transfer of information. When we are able to supersede this limit, he says, 'the "dumb" matter and mechanisms of the universe will be transformed into exquisitely sublime forms of intelligence.'

Einstein asserted that speeds greater than the speed of light have 'no possibility of existence', but can science break the old man's speed rule and help Kurzweil realise his dream? Kurzweil is optimistic that scientists will find ways for information carrying particles to travel faster than light, perhaps through wormholes. Einstein said that extremely dense objects stretch the fabric of space and time around them. If the density of an object approaches infinity, such as a black hole, that stretch can become a tear. This tear or wormhole could act as a tunnel to other parts of the universe making virtually instantaneous travel possible. The laws of physics allow wormholes but it doesn't mean that one day you would be able to travel through them.

But this has not stopped Kurzweil predicting that by 2023 computers will surpass brainpower of human and by 2045 they will surpass brainpower equivalent to that of all human brains combined, marking the end of the human era.

You will no longer be you.